Hebah Abdel-Wahab
Industrial Pharmaceutical Chemistry

Also of interest

Product-Driven Process Design.
From Molecule to Enterprise
Edwin Zondervan, Cristhian Almeida-Rivera
and Kyle Vincent Camarda, 2023
ISBN 978-3-11-101490-6, e-ISBN (PDF) 978-3-11-101569-9

Industrial Biotechnology
Mark Anthony Benvenuto, 2019
ISBN 978-3-11-053639-3, e-ISBN (PDF) 978-3-11-053662-1

Process Engineering.
Addressing the Gap between Study and Chemical Industry
Michael Kleiber, 2023
ISBN 978-3-11-102811-8, e-ISBN (PDF) 978-3-11-070949-0

Analytical Methods in Chemical Analysis.
An Introduction
Edited by: Shikha Kaushik and Banty Kumar, 2023
ISBN 978-3-11-079480-9, e-ISBN (PDF) 978-3-11-079481-6

Industrial Chemistry
Mark Anthony Benvenuto, 2023
ISBN 978-3-11-067106-3, e-ISBN (PDF) 978-3-11-067109-4

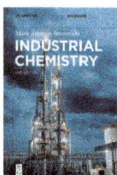

Hebah Abdel-Wahab

Industrial Pharmaceutical Chemistry

Product Quality

DE GRUYTER

Author
Dr. Hebah Abdel-Wahab
Delaware County College
901 Media Lane Road
Media, PA 19063
USA
wahabta@gmail.com

ISBN 978-3-11-131657-4
e-ISBN (PDF) 978-3-11-131686-4
e-ISBN (EPUB) 978-3-11-131725-0

Library of Congress Control Number: 2023941340

Bibliographic information published by the Deutsche Nationalbibliothek
The Deutsche Nationalbibliothek lists this publication in the Deutsche Nationalbibliografie;
detailed bibliographic data are available on the internet at http://dnb.dnb.de.

© 2024 Walter de Gruyter GmbH, Berlin/Boston
Cover image: Makidotvn/iStock/Getty Images Plus
Typesetting: Integra Software Services Pvt. Ltd.
Printing and binding: CPI books GmbH, Leck

www.degruyter.com

Preface

The book is divided into seven chapters. Chapters 1–5 discuss methods used to improve the quality of some pharmaceutical/industrial products: mouthwashes, acrylic paints, car-wash shampoos, matt and gloss paints, and some bubble-forming tablets. The methods used are based on the varying chemical properties, physical properties, and amounts of the chemical compounds used in the formulation.

Chapter 6 discusses the effect of the obstructed ability to breath by a physical block limit or by prolonged hypoventilation, or by a pulmonary condition on hypoxia, hypercapnia, and blood chemical pH imbalance. Mechanism of alternating blood pH values, methods of identifying blood chemical pH imbalance, methods of measuring P_{CO_2}, P_{O_2}, and blood pH, types of joint diseases caused due to blood chemical pH imbalance, current medicine prescribed in their treatments, and the natural treatments of such joint diseases are discussed.

Chapter 7 discusses the methods used to calculate thermodynamic parameters for simple fluorinated organic alcohols including mono and difluorinated ethanol and multifluorinated ethanol. The methods used to calculate the thermodynamic parameters, enthalpy of formations, and bond dissociation energy values are discussed. The optimized structures and Cartesian coordinates for molecules studied are included. Tabulated literature values for enthalpy of formation for molecules studied and for reference are also included.

This book is based on the actual solutions to industrial problem requests submitted to me through the Association of Consulting Chemists and Chemical Engineers, Cary, NC, USA, and the actual results of thermodynamic and kinetic independent research studies at NJIT, Newark, NJ, USA, and published articles. I hope you find the information in this book helpful to your business or project, and I also hope the information in this book increases your wealth of knowledge.

Dr. Hebah Abdel-Wahab
Edison, NJ, USA
chemistryinindustryandacademia@gmail.com

https://doi.org/10.1515/9783111316864-202

About the Author

Dr. Abdel-Wahab is a Chemical Consultant at the Association of Consulting Chemists and Chemical Engineers NC, USA, where she is consulted for independent commercial, industrial, and pharmaceutical chemical projects, and an Adjunct Professor of Chemistry at Essex County College & Middlesex County College, NJ, USA, where she has taught chemistry for allied health, preparatory chemistry, general chemistry, organic chemistry, physical chemistry, and inorganic chemistry for 17 years. Professor Abdel-Wahab earned a Ph.D. degree in chemistry from Rutgers University in 2018. She was scheduled to teach 38 different chemistry courses during her career: CHM-222, CHM-228, CHM-141, CHM-100, CHE-108, CH-112, CH-314, CHM-116, CHM-115, CHP-111, CHM-103, Chemistry-307, Chemistry-308, CHEM-4610, CE-241, CHEM-180, CHM-126, CHM-142, CHEM-1105, CHEM- 2208, CHEM- 1033, CHEM-1034, CHEM-2203, CHEM-2204, CHEM-180, CHEM-181, CH-101, CHEM-1117, CHEM-1028, CHEM-2261, CHEM-100, CHEM-111, CHEM-208, CH-132, CHE-101, CHE-102, CHEM-3184, CHEM-1083, and she adapted 115 different textbook and laboratory manuals.

Dr. Abdel-Wahab was conferred an Honorary Doctor of Science (D.Sc.) degree in Physical Chemistry in December 2022 by the International Agency for Standards and Ratings for her significant contribution in the field of chemistry and her significant contributions in the transformation of society. Prof. Abdel-Wahab was also conferred an Honorary Doctor of Science (D.Sc.) degree in Physical Chemistry in December 2022 by the British National University of Queen Mary, WP Road, London, UK.

Dr. Abdel-Wahab is the author of several book chapters in academic, commercial, industrial, and pharmaceutical chemistry. Prof. Abdel-Wahab is the author of five book chapters published by BP International: *Progress in Chemical Science Research* Vol. 1, *Current Topics on Chemistry and Biochemistry* Vol. 3, *Research Aspects in Chemical and Materials Sciences* Vol. 2, *Progress in Chemical Science Research* Vol. 4, and *Research Aspects in Chemical and Materials Sciences* Vol. 3, and the author of a book chapter published by Akinik Publications – *Research Trends in Medical Sciences* Vol. 19.

Dr. Abdel-Wahab joined the editorial cabinet as an editorial board member in 2022. She is currently an editorial board member for the *Journal of Pharmaceutics and Pharmacology Research, JPPR*, 16192 Coastal Highway, Lewes, DE 19958, USA, *International Journal of Anesthesiology and Practice*, 1603 Capital Avenue, Suite 413-A, Cheyenne, WY 82001, the USA, and *Clinical Case Reports and Trials*, Medires PUBLISHING LLC, 1603 Capital Avenue, Suite 413-A, Cheyenne, WY, 82001, USA. Dr. Abdel-Wahab is currently a reviewer for *Advances in Chemical Engineering and Science* (ACES), an SCRIP international journal, and the *Royal Society of Chemistry Journal*, RSC Burlington House, Piccadilly, London W1J 0BA, UK.

Dr. Abdel-Wahab belongs to numerous scientific and educational organizations including the International Society of Female Professionals (ISFP) and the Association of Consulting Chemists and Chemical Engineers (ACC & CE). Professor Abdel-Wahab has been published in Strathmore's *Who's Who Worldwide* (Farmingdale, NY, 2021), as an Individual Who Has Exemplified Leadership and Achievement in Their Occupation, Industry, and Profession, and she has also been published in Marquis's *Who's Who* (Berkeley Heights, NJ, 2021), as an Individual Who Demonstrates Outstanding Achievement in Their Field and who has made Innumerable Contributions to Society as a Whole.

https://doi.org/10.1515/9783111316864-203

Acknowledgments

The author wishes to acknowledge the efforts of the Association of Consulting Chemists and Chemical Engineers for making Consultants' profiles available for our clients in the USA and worldwide.

https://doi.org/10.1515/9783111316864-204

Contents

1 Chemical Formulations for Mouthwashes

1.1 Overview

Mouthwash can strengthen the teeth, prevent gum disease, kill bacteria in the mouth that causes gingivitis and improve your oral health. This chapter discusses the method used to improve the properties of the current mouthwash formulation, determine the chemical compounds and safe amounts to be used from the current formulation: distilled water, sodium chlorite, castor oil, peppermint extract, lemon juice directly from the fruit, coconut oil, sodium benzoate, citric acid, and tetrasodium EDTA, and recommend a suitable preservative for the current formulation.

1.2 Introduction

An independent business owner in New Jersey is looking to produce a small amount of homemade mouthwash to be sold and distributed locally. The ingredient included the following chemicals: distilled water, sodium chlorite, castor oil, peppermint extract, and lemon juice directly from the fruit, coconut oil, sodium benzoate, citric acid, and tetrasodium EDTA. The chemical compounds to be used in the formulation and the safe amounts of each chemical in the formulation are identified.

Mouthwash can strengthen the teeth, prevent gum disease, kill bacteria in the mouth that causes gingivitis, and improve your oral health. If the mouthwash contains fluoride, it reduces the cavity when used correctly. It is recommended by dentists to use mouthwash daily in addition to daily brushing and flossing. People with medical conditions, tooth sensitivity, dry sockets, and xerostomia find mouthwash essential in their daily routine.

Mouthwash is used to refresh and clean the oral cavity. To prevent the development of caries, a formulation containing fluorides has been used, as it destroys and inhibits the oral microbial population that generates malodors or dental plaque. Mouthwash formulation mainly consists of ethanol, water, a humectant, a surfactant, flavor, color, and an active ingredient. Ethanol is used to enhance the flavor impact and adds freshness. Humectant such as glycerin, sorbitol, or propylene glycol adds body to the formulation and improves the mouth feel when using the mouthwash. The surfactant is usually nonionic such as poly oxypropylene or poly oxyethylene copolymers; it produces foams, dissolves the flavor oil, and removes debris. The flavor is an oil used to make the mouthwash pleasant to use. The mouthwash may include an active ingredient, for example, an anticares effect or antibacterial effect.

Dental caries is a multifactorial disease produced within the dental plaque by microorganisms. It involves a process of enamel demineralization and remineralization due to the act of organic acids produced within the dental plaque by microorganisms.

https://doi.org/10.1515/9783111316864-001

Enamel demineralization is the process of dissolution of hydroxyapatite by the act of extrinsic or intrinsic acids, according to equation (1.1), leading to erosion or dental cares. Dental caries are caused by acetic or lactic acid that diffuses into the enamel pores through the plaque, decreasing the pH of the fluid surrounding the enamel crystals. To prevent dental caries, fluorides, calcium, and phosphates and antimicrobials can be used as additives in the mouthwash formulation. The presence of fluorides in the microenvironment or around the teeth inhibits demineralization and promotes the remineralization of the tooth surface. The incorporation of fluoride as fluorapatite into enamel decreases its solubility. Casein phosphopeptide amorphous calcium phosphate (CPP–ACP), has been added to mouthwash, tropical pastes, and chewing gums to increase remineralization and decrease demineralization. Antibacterial has some role in caries prevention, it helps with plaque and microbial control, chloro-hexadiene and triclosan-containing gels, toothpastes, and rinses have been used:

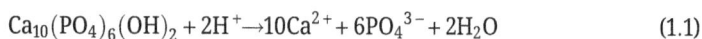

$$Ca_{10}(PO_4)_6(OH)_2 + 2H^+ \rightarrow 10Ca^{2+} + 6PO_4^{3-} + 2H_2O \qquad (1.1)$$

Mouthwash can be classified into two main types: therapeutic and cosmetic. Therapeutic mouthwash reduces plaque, bad breath, tooth decay, and gingivitis. Children under the age of 6 shouldn't be using mouthwash as they may inadvertently swallow large amounts of liquids unless directed by a dentist. Cosmetic mouthwash has no biological or chemical application beyond its temporary relief; they temporarily control bad breath and give a pleasant taste and smell.

Active ingredients used in a therapeutic mouthwash are peroxides, fluorides, chlorhexidine, essential oils, and cetyl-pyridinium chlorides. Peroxides are used in whitening mouthwash. Fluorides are used to prevent tooth decay. Chlorhexidine and essential oils are used to fight gingivitis and plaque. Cetyl-pyridinium chlorides is used to prevent bad breath.

Antimicrobial, fluoride, anesthetic, and chlorhexidine mouthwashes are examples of therapeutic mouthwashes.

Antimicrobial mouthwashes are used in the treatment of oral malodor and bad breath. Volatile sulfur compounds (VSCs) are the major factor causing bad breath; they arise from the breakdown of food, bacteria associated with dental disease, and dental plaque. Cosmetic mouthwash can also temporarily mask bad breath, but it doesn't kill the bacteria that causes bad breath and has no effect on VSCs causing bad breath. Therapeutic mouthwash containing an antimicrobial is more long term and can control bad breath. Antimicrobials in mouthwash formulations can include chlorhexidine, cetyl-pyridinium chloride, essential oils (e.g., methyl salicylate, eucalyptol, thymol, and menthol), and chlorine dioxide. Antimicrobial mouthwashes are also used in the treatment of gingivitis and plaque.

Fluoride mouthwashes are used to prevent tooth decay in children and can also be used in the treatment of xerostomia, a condition where the amount of saliva bathing the oral mucous membrane is reduced. It is found that rinsing weekly with 0.2%

NaF mouthwash or daily with a 0.02% NaF would reduce dental caries by 20–40% in children.

Anesthetic mouthwashes are used to relieve pain, and they contain anesthetics, dyclonine hydrochloride, phenol, lidocaine, benzocaine, tetracaine hydrochloride, or butamin.

Whitening mouthwash mostly contains 1.5–2% hydrogen peroxide or 10% carbamide peroxide.

Chlorhexidine mouthwash, without the use of antibiotics, is found to be effective for AO prevention following extractions. Dry socket, alveolar osteitis, is a condition that might follow an extraction procedure and intense pain 2–3 days after the procedure.

Additives that make a therapeutic mouthwash can be classified into either antibacterial agents or compounds that effect VSC formation. Antibacterial agents help reduce the number of anaerobic bacteria in the mouth. Examples of antibacterial agents that are used in mouthwash formulation are cetyl pyridinium chloride, essential oils/antiseptic formulations, triclosan, and chlorhexidine gluconate. It is found that the use of 0.75% cetyl pyridinium chloride mouth rinse reduces plaque by 35% over a 6-month period, and the use of 0.2% chlorhexidine gluconate oral rinse solution causes reduction in VSC level (it shows some side effects and alters the sense of taste and causes staining of mouth and tongue).

Compounds that effect VSC formation neutralize VSCs to improve quality of breath. Examples of ingredients that effect VSCs are zinc compounds and chlorine dioxide (sodium chlorite is an oxidizing agent and is able to degrade sulfur-containing amino acids, which is the building block of VSCs, making fewer of them available to bacteria, leading to fewer VSCs formed). It is found that using mouthwash containing 1% sodium chlorite reduces VSC for a period of 8 h. Some metal ions have the ability to oxidize the thiol, sulfur-based molecules found on VSCs. Examples of metal ions are zinc, copper, magnesium, tin, and sodium; zinc also has an antibacterial property, doesn't cause tooth staining like other metal ions, and has low toxicity. A 1% zinc acetate mouthwash has a significant effect on VSC levels, even 3 h after use. A typical formulation for essential oils/antiseptic mouthwash is 0.092% eucalyptol, 0.042% menthol, 0.06% methyl salicylate, and 0.064% thymol in a solution containing 21.6% to 26.9% alcohol, high concentration of alcohol in the formulation (25% ethanol) may cause drying effect on oral tissue, increasing breath malodor. The most effective mouthwash for bad breath is found to use a multifaceted approach using a combination of formulations. Other products used for breath control may include gum, toothpaste, mints, drops, lozenges, and sprays. It is found that six compounds are mainly used in the formulation of mouthwash designed to fight bad breath (halitosis); zinc chloride, triclosan, essential oils/antiseptic formulations, chlorhexidine gluconate, cetyl-pyridinium, and chlorine dioxide. Studies show that using a 0.09% zinc chloride mouth rinse is effective in reducing the formation of calculus in people.

Sodium chlorite is the main ingredient used in the newer classes of mouthwashes such as ProFresh, CloSYS, TheraBreath, and Oxyfresh. It is sometimes used as a water purifier. These mouthwashes are said to freshen breath up to 6 h. SmartMouth uses sodium chlorite as a main ingredient, but it must be mixed with zinc chloride right before use. The zinc ions block receptor sites of amino acids after eating food containing amino acids, so the bacteria can't produce rancid gases, and it lasts for up to 12 h before another rinse is needed.

Flavors are one of the main ingredients that make a mouthwash, as it gives flavor to the formulation and makes the mouthwash pleasant to use. The different known flavors that can be used in a mouthwash ingredient are: clove leaf oil, cubeb oil, cedarwood oil, eucalyptus oil, lemon oil, Italian type, coldpressed, myrrh oil, mentha arvensis oil, sweet orange oil, peppermint oil, sucralose, saccharin sodium, dihydrate, granular, spearmint oil, camphor, methyl salicylate, and cinnamon extract.

1.3 Experimental

The chemical composition of the ingredients used in the making of the mouthwash is identified in Tab. 1.1.

Distilled water (purified water) is used as a solvent in which the ingredients are soluble in. It is a cooled boiling water steam returned to its liquid state after removing 99.9% of the minerals dissolved in water.

Sodium chlorite is used as an antimicrobial agent in mouthwash. It reduces bacteria living in the mouth causing bad breath. It also degrades building blocks of the VSCs causing bad breath, and it degrades sulfur-based amino acids, causing fewer of them being available to bacteria, and fewer VSCs are formed. It is found that one-time use of mouthwash containing 1% sodium chlorite is effective in reducing the RVS level for 8 h and beyond.

Castor oil has the ability to break biofilm, a protective coating created by detrimental bacteria in the microbiome. It makes the environment more hospital to healthy bacteria and eradicates the bad bacteria. It also reduces inflammation of the gums [25]. Castor oil is a slightly toxic chemical, and a safe amount of castor oil is found to be below 5 g/kg, below mass per cent (m/m) of 0.099%.

Peppermint extract is used as a flavoring agent, antiseptic, and antiviral agent. There is no evidence that it has the ability to treat any medical condition. However, the high concentration of menthol in the extract causes side effects in children and infants when annihilated.

Sodium benzoate is used as a preservative in food, medicine, and cosmetics. It is converted to benzoic acid under these conditions, which are fungistatic and bacteriostatic. Due to the insolubility of benzoic acid in the water, it isn't used directly as a preservative. Its concentration as a preservative is determined as 0.1% by the FDA.

Lemon juice is a good source of citric acid, containing 1.44 g/oz citric acid, which is 4.83% by mass citric acid in lemon juice [29]. Citric acid is known to cause tooth enamel to dissolve quickly. More dentin is exposed as enamel dissolves, causing teeth cracks and chips, and the edges of the teeth to become more irregular.

Tetrasodium EDTA is a water-soluble acid with chelating properties and is an emulsion stabilizer. It bonds with metal ions in solution, causing them to be inactive, and is used as a preservative for cosmetic formulation and skin care products, creams, lotions, etc. It prevents the change in pH, texture, and color of skincare products. And it is used as a copreservative in skincare formulations. Also, it can enhance foaming and cleaning ability when binding to iron, calcium, or magnesium.

Tab. 1.1: Structures, molecular formulas, and molecular weights of all household chemical compounds used in the current formulation.

Household Chemical and Chemical Name	Chemical Structure	Molecular Formula	Molecular Weight (g/mol)
Distilled water		H2O	18.015
Sodium chlorite		NaClO2	90.44
Castor oil	Major component of castor oil, triester of glycerol and ricinoleic acid	C57H10409	933.4

Tab. 1.1 (continued)

Household Chemical and Chemical Name	Chemical Structure	Molecular Formula	Molecular Weight (g/mol)
Peppermint extract/Oil	(Menthol), (3,7-Dimethyl-l-oxaspiro[3.5]nonane), (Menthyl acetale), (p-Menthan-3-one) and (Menthofuran)	C62H10807	965.5
Sodium benzoate		C7H5NaO2	144.11
Lemon Juice (Citric acid)		$C_6H_8O_7$	192.124

Tab. 1.1 (continued)

Household Chemical and Chemical Name	Chemical Structure	Molecular Formula	Molecular Weight (g/mol)
Tetrasodium EDTA		C10H12N2Na4O8	380.171
Lemongrass oil	 (Methyleugenol), (L-Borneol), (Nerol), (Dihydromyrcene) and (Neral)	C51H84O5	777.2

Coconut oil contains 80–90% unsaturated fat, is 100% fat, and contains a trace of minerals and vitamins. It is used for cooking food, as a fuel source in industry, and as a base ingredient for the manufacture of soap.

Lemongrass oil is known to be used as an herbal alternative (0.25%) to chlorhexidine (0.2%) in mouthwash. It prevents and treats periodontal diseases.

1.4 Conclusion

The therapeutic mouthwash formulation shouldn't include the following chemical compounds: citric acid, lemon juice, EDTA, and coconut oil. Citric acid/lemon juice is known to weaken and dissolve tooth enamel, and both EDTA and coconut oil are known neither to be used in the pharmaceutical industry nor in mouthwash formulations. EDTA is commonly used in skin products as a chelating agent, and coconut oil is a fat that is known to be used in cooking, fuel, and in the making of soaps.

The therapeutic mouthwash formulation may include 1.0% Sodium chlorite as an active ingredient, 0.1% sodium benzoate as a preservative, water as the solvent, and less than 0.09% by mass castor oil. To improve the quality of the mouthwash formulation, other chemical compounds may be used: glycerin as a humectant and poly oxypropylene co-polymer as a surfactant. Color and flavor may be used to improve the quality of the mouthwash. Some other additives, such as zinc chloride, may also be used; it Is known to freshen up breath 12 times higher when combined with sodium chlorite.

The recommended preservative for mouthwash is sodium benzoate. The safe amount of hydrogenated castor oil is below 5 g/kg.

Chapter Questions

1) What's the aim of the chapter?
2) What's the chemical composition of peppermint extract?
3) What are types of mouthwashes?
4) What are the chemicals used in the current formulation? Which chemicals are found to be harmful to the current formulation? What is the preservative recommended and its safe amount?
5) Write the equation for enamel demineralization? And methods to prevent tooth decay?

References

[1] https://www.westenddental.com/blog/is-mouthwash-necessary/
[2] Johansson I, Somasundaran P, Handbook for Cleaning/ Decontamination of Surface, Elsavier, Kidlington, Oxford OX5, 2007. Pp 382–383.
[3] Manton David J., Hayes-Cameron L., Handbook of Pediatric Dentistry 4th edition, 2013, Dental Caries, Mosby Elsevier, Canberra Australia. Pp. 49–78.
[4] https://www.ada.org/resources/research/science-and-research-institute/oral-health-topics/mouth rinse-mouthwash
[5] Torres CR, Perote LC, Gutierrez NC, Pucci CR, Borges AB. Efficacy of mouth rinses and toothpaste on tooth whitening. Oper Dent 2013;38(1):57–62.

[6]   Marinho VC, Higgins JP, Logan S, Sheiham A. Topical fluoride (toothpastes, mouthrinses, gels or varnishes) for preventing dental caries in children and adolescents. Cochrane Database Syst Rev 2003(4):CD002782.

[7]   Hasson H, Ismail AI, Neiva G. Home-based chemically induced whitening of teeth in adults. Cochrane Database Syst Rev 2006(4):CD006202.

[8]   Sharma N, Charles CH, Lynch MC, et al. Adjunctive benefit of an essential oil-containing mouthrinse in reducing plaque and gingivitis in patients who brush and floss regularly: a six-month study. J Am Dent Assoc 2004;135(4):496–504

[9]   Blom T, Slot DE, Quirynen M, Van der Weijden GA. The effect of mouthrinses on oral malodor: a systematic review. Int J Dent Hyg 2012;10(3):209–22.

[10]   Araujo MW, Charles CA, Weinstein RB, et al. Meta-analysis of the effect of an essential oil-containing mouthrinse on gingivitis and plaque. J Am Dent Assoc 2015;146(8):610–22.

[11]   Sharma N, Charles CH, Lynch MC, et al. Adjunctive benefit of an essential oil-containing mouthrinse in reducing plaque and gingivitis in patients who brush and floss regularly: a six-month study. J Am Dent Assoc 2004;135(4):496–504

[12]   Fedorowicz Z, Aljufairi H, Nasser M, Outhouse TL, Pedrazzi V. Mouthrinses for the treatment of halitosis. Cochrane Database Syst Rev 2008(4):CD006701.

[13]   Leary, Kecia S., Nowak, Arthur J., Pediatric dentistry: infancy through adolescence, Prevention of Dental Disease, P455–460; 2019.

[14]   Mariotti AJ, Burrell, K.H. Mouthrinses and Dentifrices. 5th ed. Chicago: American Dental Association and Physician's Desk Reference, Inc.; 2009.

[15]   Mariotti AJ, Burrell, K.H. Mouthrinses and Dentifrices. 5th ed. Chicago: American Dental Association and Physician's Desk Reference, Inc.; 2009.

[16]   Hasson H, Ismail AI, Neiva G. Home-based chemically-induced whitening of teeth in adults. Cochrane Database Syst Rev 2006(4):CD006202.

[17]   Kerr AR, Corby PM, Kalliontzi K, McGuire JA, Charles CA. Comparison of two mouthrinses in relation to salivary flow and perceived dryness. Oral Surg Oral Med Oral Pathol Oral Radiol 2015;119 (1):59–64.

[18]   https://www.animated-teeth.com/bad_breath/t5_halitosis_cures.htm

[19]   https://www.webmd.com/diet/distilled-water-overview#1

[20]   https://www.animated-teeth.com/bad_breath/t5_halitosis_cures.htm

[21]   https://pubchem.ncbi.nlm.nih.gov/compound/Castor-oil

[22]   Andrade IM, Andrade KM, Pisani MX, et al. Trial of an experimental castor oil solution for cleaning dentures. Braz Dent J. 2014;25(1):43–47.

[23]   Badaró MM, Salles MM, Leite VMF, et al. Clinical trial for evaluation of Ricinus communis and sodium hypochlorite as denture cleanser. J Appl Oral Sci. 2017;25(3):324–334.

[24]   Salles MM, Badaró MM, Arruda CN, et al. Antimicrobial activity of complete denture cleanser solutions based on sodium hypochlorite and Ricinus communis – a randomized clinical study. J Appl Oral Sci. 2015;23(6):637–642.

[25]   Vieira C, Evangelista S, Cirillo R, et al. Effect of ricinoleic acid in acute and subchronic experimental models of inflammation. Mediators Inflamm. 2000;9(5):223–228.

[26]   Gosselin, R.E., H.C. Hodge, R.P. Smith, and M.N. Gleason. Clinical Toxicology of Commercial Products. 4th ed. Baltimore: Williams and Wilkins, 1976., p. II-152.

[27]   https://en.wikipedia.org/wiki/Peppermint_extract

[28]   https://en.wikipedia.org/wiki/Sodium_benzoate

[29]   Penniston K.L., Nakada S.Y., Holmes R.P., Assimos D.G., Quantitative Assessment of Citric Acid in Lemon Juice, Lime Juice, and Commercially-Available Fruit Juice Products, National Library of Medicine, Journal of Endourology, 2008;22(3):567–70.

[30]   https://pubchem.ncbi.nlm.nih.gov/compound/Citric-acid
[31]   https://www.glowdental.co.uk/how-citric-acid-affects-your-teeth/
[32]   https://en.wikipedia.org/wiki/Tetrasodium_EDTA
[33]   https://www.lorealparisusa.com/ingredient-library/disodium-edta
[34]   https://www.hsph.harvard.edu/nutritionsource/food-features/coconut-oil/
[35]   https://www.ncbi.nlm.nih.gov/pmc/articles/PMC4625327/
[36]   https://pubchem.ncbi.nlm.nih.gov/compound/Water
[37]   https://en.wikipedia.org/wiki/Sodium_chlorite
[38]   https://en.wikipedia.org/wiki/Castor_oil
[39]   https://pubchem.ncbi.nlm.nih.gov/compound/Peppermint-oil
[40]   https://pubchem.ncbi.nlm.nih.gov/compound/Sodium-benzoate
[41]   https://pubchem.ncbi.nlm.nih.gov/compound/Castor-oil
[42]   https://pubchem.ncbi.nlm.nih.gov/compound/Lemongrass-oil
[43]   Charles C.H, Cronin M. J., Dembling W. Z., Petrone D. M., McGuire J. A., Anticalculus efficacy of an antiseptic mouthrinse containing zinc chloride, Journal of American Dental Association, 2001;132 (1):94–8, doi: 10.14219/jada.archive.2001.0033.
[44]   https://usatoday30.usatoday.com/news/health/2008-05-13-bad-breath_N.htm
[45]   http://dentalingredients.spectrumchemical.com/products/oral-care-flavorants

2 Chemical Formulations for Acrylic Matt and Acrylic Gloss Paints

2.1 Overview

Paints are used to prolong the life of natural and synthetic materials and to protect it; it acts as a barrier against environmental conditions. Mainly paints are made up of solvent, binder, extenders, pigment(s), and some additives. A solvent can be either organic or inorganic, and it is used as a medium to dissolve paint contents together and make them uniform. It can also be used as a thinner. Pigments are used to give color to the paint. Binders are used to hold pigments in place. Extenders have large pigment particles to improve the adhesion properties of the paint, and they are also used to strengthen the film and save the binder. This chapter discusses the methods used to improve the properties of acrylic matt paint and methods used to increase the scrub resistance test of the current formulation. This chapter also presents the types of additives and chemical compounds that would potentially increase the quality and increase the scrub resistance test for acrylic paints.

2.2 Introduction

Professional Specialty Chemicals Factory, a polymer production unit in Saudi Arabia, are looking to improve its formulation for acrylic paints to give a better scrub resistance test after decades of using the same formulation. The current formulation consists of the following chemical compounds: 0.4% natrosol 250-HHBR, 44.39% water, 0.14% sodium bicarbonate, 2.0% polyoxyethylene 25 octyl phenol, 2.0% octyl phenol polyglycol ether sulfate sodium salt, 0.5%, provichem, 2.0% butyl acrylate, 47% vinyl acetate, 0.1258% potassium persulfate, 0.0234% tertiary butyl hydrogen peroxide, 0.06% hydrogen peroxide, 0.05% sodium formaldehyde sulfoxylate, 1.0% dibutyl phthalate, 0.1% formaline, 0.05% biocide, 0.1% Silquest A-171, and 0.06% defoamer. The scrub resistance test of the current formulation shows 60% effectiveness compared to the reference; the coat breaks and gloss change were observed after 60 cycles for the current formulation compared to 100 cycles for the reference.

Paints are used to prolong the life of natural and synthetic materials and to protect it; it acts as a barrier against environmental conditions. Paints contain pigment (s), binder, extender, solvent, and some additives. The pigment is used to give color and opacity. Binders are matrices and are used to hold the pigment in place. Extenders have larger pigment particles to improve adhesion and to strengthen the film and save the binder. A solvent can be either an organic solvent or water and is used as a thinner and to dissolve paint components and make it uniform. Additives are commonly used to improve the properties of the paint. *Contents of an acrylic white matt*

https://doi.org/10.1515/9783111316864-002

emulsion paint are known as follows: pigments 25%, extender pigments 12%, additive 5%, solvent 44%, and binder 14%.

To control paint gloss, extender pigments are used. Extender pigments give extra weight to the paint and are low-cost pigments. There are two types of pigments: extender pigments and prime pigments. Prime pigments are those that contribute to both dry and wet hide in paint. There are two main types of prime pigments: inorganic and organic prime pigments. Inorganic prime pigments are earthy colors, duller and more durable for exterior paint application. Examples are yellow ochre, umber, and red oxide. TiO_2, titanium dioxide, provides the actual color within the container. Organic prime pigments aren't very durable for exterior paint application and provide brighter colors; examples are phthalo blue and hansa yellow. An extender pigment is a naturally occurring chemical substance (usually having a white color) that is added to paint or coating to improve its properties, such as durability, cost, and resistance to corrosion or wear.

Ingredients that provide a binding effect that holds the pigments together as dry film after the liquid solvent evaporates are called binders. Paint binders directly relate to paint performance, including gloss retention or fade resistance, washability, adhesion, and scrub resistance. The binder in many emulsion paints is based on homopolymers or copolymers of vinyl acetate and a propenoate (acrylic) ester. Other acrylic esters used as comonomers with ethenyl ethanoate are ethyl propenoate, butyl propenoates, or a copolymer of butyl propenoate and methyl 2-methylpropenoate. Many commercial extenders are glycerin-based. In fact, glycerin is already an ingredient in most acrylic paints. It gives the paint viscosity in addition to extending the drying time. We can heighten this effect by using glycerol mixed with water. Propylene glycol is also used for thicker applications.

Binders in various emulsion paints are based on copolymers or homopolymers of a propenoate (acrylic) ester and ethenyl ethanoate (vinyl acetate)

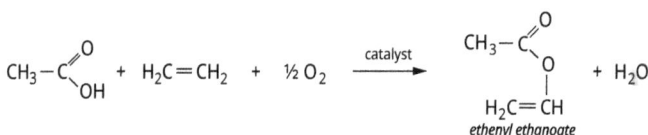

Fig. 2.1: Homopolymers of an acrylic ester and vinyl acetate.

Solvents are liquids that allow the paint to be applied to the surface directly from the containers. Solvents allow the uniform combination of binders, considered solids, and pigments with the liquid solvent. Types of liquid solvents depend on the type of paint. For alkyd paints, the liquid solvent is thinner and is typically another type of solvent; for latex paints, water is the main liquid solvent. More solids in the ingredients allow for higher quality paints, 35–40%. How much paint will remain on the surface after the liquid solvent is evaporated depends on the amount of solid in the paint formulation.

The make-up of the solids effects the overall paint performance. High solids by volume don't always give a high-quality product.

To modify the properties of paint and create additional performance properties, additives are used. The percent of additives in the formulation is known to be 0.1–1%. Additives may include algaecides, bactericides, anti-settling agents, driers, thixotropic agents, dispersants, silicones, and cross-linkers. The molar mass of a polymer is greatly increased by using a cross-linker, silicates, causing the polymer to become a three-dimensional molecule and form a hard film to become resistant to chemical and environmental conditions. Algaecides are used to protect exterior paint films from being disfigured from lichen, algae, and molds, and bactericides are used to preserve water-based paints in their containers. To prevent pigment settling, antisettling agents (surfactants) may be used. To accelerate drying time, driers are used. Thixotropic agents are known to be used to create a jelly-like texture of the paint that would break down to liquid when the paintbrush is placed in it or when stirred. To improve weather resistance, silicones are used. To stabilize and separate pigment particles and prevent them from aggregating, dispersants are used.

Water-borne emulsion paints are better for the environment than paints made with organic liquid solvents. Emulsion paints are made using emulsion polymerization, and monomers are dispersed in water as an emulsion. The molecular weight of the polymers produced is 500,000–1,000,000 a.m.u. For a better application of paint, a solvent thinner can be used, and it is either water or organic solvent.

An ideal paint has high scratch and heat resistance, high color stability against ultraviolet and visible light, high water and corrosion resistance, high opacity, a relatively quick drying time and is easily applied to the surface, has high durability and flexibility, has a good flow out of the application, and forms a strong protective film.

Surfactants, surface active agents, are compounds that lower the surface tension of liquid or interfacial tension between solid and liquid making them more soluble. Stabilizing effects for latex paints are caused by using a soluble surfactant, as they can form micelles and aggregate structures in solutions. In paints, surfactants are used as an emulsifier. Surfactants have a positive and negative role in being used in a paint formulation. The positive effect would be stabilizing the dispersion of polymer molecules during emulsion polymerization, improving the mechanical stability of paints, and allowing the paint to coat the surface more easily. The negative effect would be decreasing the resistance of the coating as it can be washed out of the coating and enter the environment; some of these surfactants are toxic to the environment, problems with adhesion, and loss of optical clarity. Other chemicals in paint can alter the overall effect of surfactants on paint. The amount of TiO_2 affects the elasticity in latex paint. Chains that consist of alternating silicon and oxygen atoms and siloxane chains with siloxane tails in polymers have been found to resist hydrolysis and prevent the breakdown of polymer chains; breakdown of polymer chains can cause cracking in the paint film. Siloxane is used in soaps, cosmetics, defoamers, and deodorants.

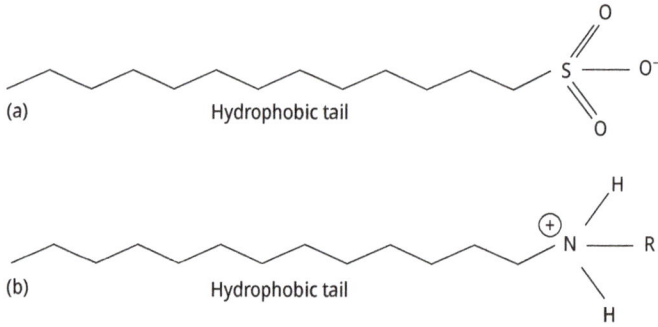

(a) Hydrophobic tail

(b) Hydrophobic tail

Fig. 2.2: (a) and (b) are the general formula for a surfactant.

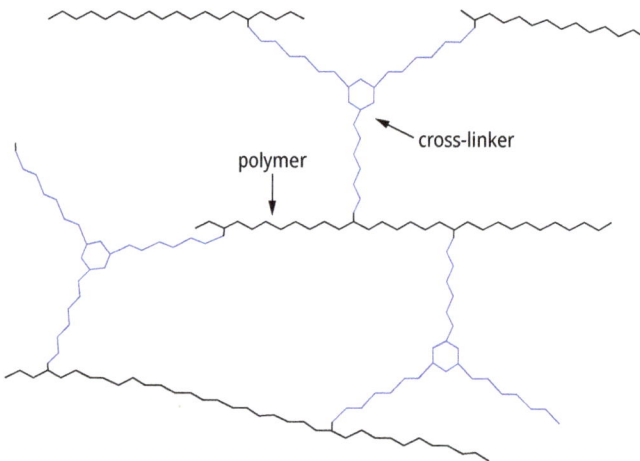

cross-linker

polymer

Fig. 2.3: A cross-linker.

Polymers are high molecular substances that are made up of small low-molecular weight chemical species held together by a chemical bond called monomers. Polymers are used in coatings, polyurethane, and plastics, and they are used as a binder in paints. The first polymer studied by chemists is natural or biological polymers such as protein, starch, and cellulose. An example of a polymer would be polyethylene consisting of ethylene monomers that are chemically bonded.

Polyethylene is a polymer consisting of thousands of ethylene units. The name polyethylene is a name arousing from the prefix poly, which means "many," and the monomer name (ethylene). Synthetic polymers can be either condensation or addition polymers based on the method used in synthesizing them. In addition to polymerization, many molecules are linked together by an addition reaction. The monomer molecules must be unsaturated, containing multiple bonds that can undergo addition reactions. In a free radical addition, an initiator is used to prepare an addition polymer. An initiator is a compound that can produce free radicals, and organic compounds with O–O group (peroxides) are commonly used as initiators. When organic peroxides are heated RO–OR, the O–O bond breaks and free radical form (species having unpaired electrons). Alkene monomers would react with the free radical formed and produce another free radical, which in turn reacts with another monomer molecule and produce another free radical molecule in a chain reaction. The reaction terminates when a free radical at the end of the chain reacts with another free radical. In condensation polymerization, many monomer molecules are linked together, and a polymer is formed by a condensation reaction, where a water molecule is removed by joining two molecules. Examples of condensation polymers are polyesters and polyamides. The repeating monomer units are joined together by an ester group in polyesters and by an amide group in polyamides.

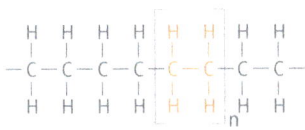

Fig. 2.4: Shown is a piece of polyethylene.

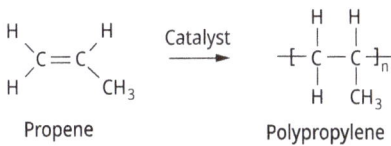

Propene   Polypropylene

Fig. 2.5: Example for addition polymerization of propene into polypropylene.

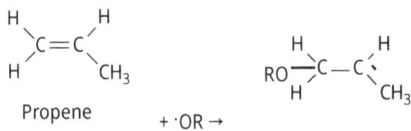

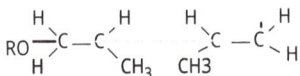

$$RO–OR \rightarrow RO^{\cdot} + {^{\cdot}OR}$$

Fig. 2.6: Example for free radical polymerization of propene using organic peroxide.

$n\text{HOH}_2\text{C} - \text{CH}_2\text{OH} \quad + \quad n\text{HOOC} -\langle\rangle- \text{COOH}$

Ethylene glycol   Terephthalic acid

$\left[\text{OCH}_2\text{CH}_2 - \text{O} - \underset{\text{O}}{\overset{\text{O}}{\text{C}}} -\langle\text{O}\rangle- \overset{\text{O}}{\underset{\text{C}}{}} \right]_n$

Terylene or dacron

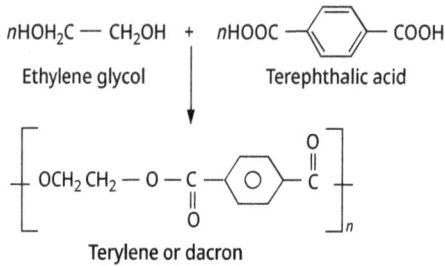

Fig. 2.7: Example of polyester Dacron formed from condensation polymerization of ethylene glycol and terephthalic acid.

The technological and physicomechanical properties of various polymers are dependent on the branching that occurs in the polymer chain compared to a linear polymer. An increased degree of branching in the polymer leads to a decrease in strength properties. For example, branched polybutadiene, divinyl styrene, and other branched elastomers exhibit low elasticity and breaking strength.

Polymers are classified based on their composition into homopolymer and copolymer. A homopolymer is a polymer that consists of only one monomer. A copolymer can be either alternate, random, or block copolymer. An alternate copolymer consists of alternating monomers ab–ab–ab, a block copolymer consists mainly of two or more monomer chains combined, aaaaa–bbbbb, and a random copolymer consists of monomers combined in a random array aa–bb–bb–ab–ba.

The durability of concrete structures can be increased by the application of protective coatings, such as paints.

Protective coatings can be used for the modernization of older concrete works, rehabilitation of deteriorated concrete, and for restoration of its original properties. The chemical composition of the protective coatings determines the film properties. Its properties can be modified by altering the amounts of additives and pigments. Pigments affect the permeability and strength of the film and control the color and gloss.

Latex paints have 10 times greater water diffusion, and it is less homogenous than solvent paint films.

When concrete is exposed to the environment, it deteriorates, especially when exposed to aggressive environments: industrial and coastal areas. The performance of paints depends on the chemical composition of their formulation. The better protection against water penetration was greater with increasing the amounts of binders (resin) and decreasing amounts of polyvinyl chloride (PVC) polymer. In case of concrete with low compressive strength, paints are shown to be efficient for its protection, while high-compressive strength concrete paints have smaller importance as it already has a low permeability. Paints are considered for visual aspects with concrete of high compressive strength. Ref [30]. Decreasing amounts of PVC in the paint formulation from 75% to 0% showed a decrease in water absorption and an increase in coating effectiveness from 6% to 91%.

2.3 Experimental

The paint is prepared by emulsion polymerization, where soap, free radical catalysts (potassium persulfate, peroxides), and monomers are emulsified in water. The water-soluble free radical catalyst is added to induce polymerization.

Natrosol 250-HHBR is used as a nonionic water thickener, a protective colloid, suspending agent, and stabilizer and gives chemical mechanical stability and controls rheology after, during, and before application.

Water is used as a dispensing solvent for emulsion polymerization.

Sodium bicarbonate is used as a pigment, and it dissolves immediately upon addition to the reaction mixture; a thick textured effect is the result. If more baking soda is used, it causes the paint to fluff up and become more textured.

Polyoxyethylene 25 octyl phenol is used as a spermaticide. Octyl phenol polyglycol ether sulfate sodium salt is used as a surfactant and a buffer in emulsion polymerization, copolymerization, and homo-polymerization of monomers, and the sodium salt is used to main pH 6.5–7.5.

Octylphenol polyglycol ether sulfate sodium salt is used in the emulsion polymerization of acrylate and methacrylate esters, styrene, and vinyl esters. Sodium salt can be used in homopolymerization as well as in copolymerization of these monomers; the sodium salt maintains a pH level ranging from 6.5 to 7.5.

Provichem, sodium vinyl sulfonate is a surfactant in emulsion polymerization; it produces a copolymer with high water stability.

Butyl acrylate is used as a monomer in copolymerization.

Vinyl acetate is used as the other monomer in homopolymerization or copolymerization.

Potassium persulfate is used as an initiator for free radical polymerization and is also known to be used as a bleaching agent.

Tertiary butyl hydrogen peroxide is commonly used as an oxidizing catalyst.

Hydrogen peroxide is known to be used as an initiator in vinyl polymerization in a homogeneous system.

Sodium formaldehyde sulfoxylate is used in redox-reaction polymerization as a reducing agent and as a release and stripping agent in the textile industry; it is also used in emulsion polymerization as a reducing agent, and it can be used in combination with all other oxidizing agents, and the recommended usage would be within the range 0.05–0.2% for postpolymerization reactions, and it can be dry-stored for 12 months at 25 °C.

Dibutyl phthalate is used in adhesives, printing inks, and lacquers as a plasticizer, and it can be used with other phthalates for PVC compounds as a secondary plasticizer.

Formalin is used as a preservative in some types of adhesives, such as vinyl, and it may result as a by-product from some polymerization reactions. *Biocide* is known to be a chemical substance that has a controlling effect on any harmful organism by biological or chemical means.

Tab. 2.1: Lists the chemical name, molecular formula, and molecular weight of chemical compounds used in the current acrylic paint formulation.

Chemical Name	Percent (%) Used in the Current Formulation	Molecular Weight (g/mol)	Chemical Structure	Molecular Formula
Water	44.09	18.902		H2O
Natrosol 250- HHBR	0.4	806.9	N/A	$C_{36}H_{70}O_{19}$
Sodium Bicarbonate	0.14	84.007		$NaHCO_3$
Polyoxyethylene (25) octyl phenyl ether	2.0	250.379	N/A	(C2-H4-O)mult-C14H22-O
Octyl phenol polyglycol ether sulfate sodium salt	2.0	Buffer and Surfactant	N/A	Octyl phenol polyglycol ether sulfate sodium salt
Provichem (Sodium vinylsulfonate)	0.5	130.10		$C_2H_3NaO_3S$
Buyl acrylate	2.0	128.17		$C_7H_{12}O_2$

Name	Amount	MW	Structure	Formula
Vinyl acetate	47.0	86.09		$C_4H_6O_2$
Potassium Persulfate	0.1258	270.33		
Tertiary butyl hgrogen peroxide	0.0234	90.12		$C_4H_{10}O_2$
Hydrogen peroxide	0.06	34.0147		H_2O_2
Sodium formaldehyde sulfoxylate	0.05	154.12		CH_7NaO_5S

(continued)

Tab. 2.1 (continued)

Chemical Name	Percent (%) Used in the Current Formulation	Molecular Weight (g/mol)	Chemical Structure	Molecular Formula
Di butyl phthalate	1.0	278.34		$C_{16}H_{22}O_4$
Formaline	0.1	30.031		CH_2O
Biocide	0.05	biocide	N/A	Biocide
Silquesit A-171 (Vinyltrimethoxysilane)	0.1	148.23		$C_5H_{12}O_3Si$
Defoamer	0.06	Defoamer	N/A	Defoamer
Total Percentage	100	N/A	N/A	N/A

Additives are chemical substances added in small amounts to the main ingredient to improve some chemical or physical properties of the material being synthesized. Silquesit A-171, vinyltrimethoxysilane, offers silanes and vinyl compounds some functionality by cross-linking organic polymers; the cross-linked Si–O–Si bond formed is highly resistant to UV light, other chemicals, and exposure to moisture. It can also be added as a monomer in emulsion polymerization to form modified silane latexes, functioning as a cross-linker forming stable Si–O–Si bond linkages.

2.4 Discussion

The current formulation consists of the following chemical compounds: 0.4% natrosol 250-HHBR, 44.39% water, 0.14% sodium bicarbonate, 2.0% polyoxyethylene 25 octyl phenol, 2.0% octyl phenol polyglycol ether sulfate sodium salt, 0.5%, provichem, 2.0% butyl acrylate, 47% vinyl acetate, 0.1258% potassium persulfate, 0.0234% tertiary butyl hydrogen peroxide, 0.06% hydrogen peroxide, 0.05% sodium formaldehyde sulfoxylate, 1.0% di butyl phthalate, 0.1% formaline, 0.05% *biocide* , 0.1% Silquest A-171, and 0.06% defoamer.

The current acrylic paint formulation consists mainly of 44.39% water as solvent, 49% pre-emulsion monomers (vinyl acetate and butyl acrylate), and 6.61% additives (natrosol 250-HHBR, sodium bicarbonate, polyoxyethylene 25 octyl phenol, octyl phenol polyglycol ether sulfate sodium salt, Provichem, 2.0% butyl acrylate, potassium persulfate, tertiary butyl hydrogen peroxide, hydrogen peroxide, sodium formaldehyde sulfoxylate, dibutyl phthalate, formaline, *biocide*, Silquest A-171, and defoamer).

The current percentage of each component in the current formulation provided are pigments (sodium bicarbonate) 0.14%, extenders (none used), additives 6.4629%, solvent 44.39%, and binders 49%.

The main pre-emulsion monomer used is of only one type of monomer, and vinyl acetate and butyl acrylate are only of 2% and is used as additive in the current formulation. The binder formed from the pre-emulsion polymerization of the monomers used in the formulation would mainly be a homopolymer consisting of vinyl acetate monomers and some butyl acrylate impurities.

The amount of surfactant used as an additive is 4.5%, which is very high and causes a lower scrub resistance test for the current formulation. The addition of surfactants does not always have a positive effect on all properties. The water resistance of the coating can be decreased with surfactant addition since surfactants can be very water-soluble and will easily wash out of a coating. This problem of moisture resistance is a particularly prevalent problem for art conservation as well as problems with adhesion, loss of optical clarity, and dirt pickup caused by polyether surfactants in contemporary acrylic emulsion used in artworks bearing acrylic coats. While the type and amount of surfactant determine what properties will be affected, other chemicals in paint can alter the overall effect the surfactants may have on the paint. Elasticity has been found to either increase or decrease in latex paints depending on the amount of TiO_2 present.

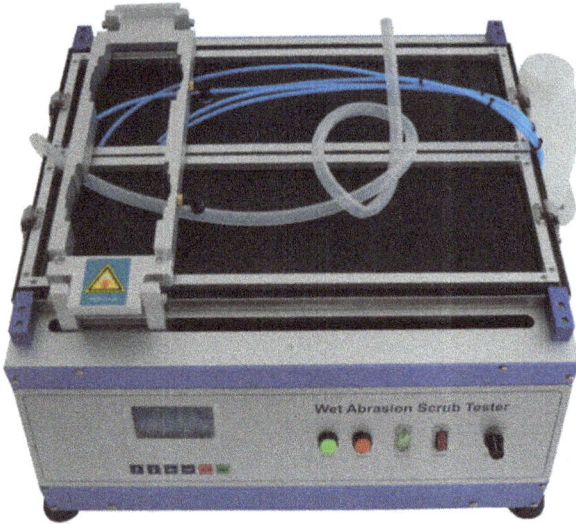

Fig. 2.8: Sheen scrub tester used in the scrub resistance test for the current acrylic paint formulation.

2.5 Conclusion

In order to improve the quality of the paint and increase its scrub resistance, missing chemical compounds from the current formulation, pigments, extender, and another type of pre-emulsion monomer must be added to the formulation. The use of pigments and extenders in the formulation would form a firm protective layer and prevent the dissolution of the paint film from the surface and to improve the scrub resistance test of the acrylic paint, and it is recommended to avoid using copious amounts of surfactants.

The current formulation consists of pigments 0.14% in the form of sodium bicarbonate compared to the literature amount 25%. Extender pigments weren't used in the current formulation compared to the literature amount 12%, surfactants 4.5% compared to the literature amount 0–1%, and binders 49% compared to the literature amount 14%. The current paint scrub resistance test shows that the paint operates with 60% effectiveness compared to the reference due to missing 40% of the main ingredients that make good acrylic paint.

Copious amounts of surfactants 4.5% (2.0% octyl phenol polyglycol ether sulfate sodium salt, 2.0% octyl phenol polyglycol ether sulfate sodium salt, and 0.5% Provichem) have been causing a low-scrub resistance test of current formulation compared to the reference. The large amounts of surfactants are used as an additive, 4.5%, has a negative effect on the formulation, causing a low-scrub resistance test result compared to the reference.

Chapter Questions

1) What's the aim of the chapter?
2) What's the main chemical composition of paints and its function in the formulation?
3) What's the chemical name for Provichem?
4) What are the chemicals used in the current formulation? Which chemicals are found to be harmful to the current formulation? And what are the recommended additives to increase its quality?
5) What's a crosslinker and its function?

References

[1]   http://pscf.com.sa/
[2]   The Essential Chemical Industry, 2013, Paints. Retrieved February 9, 2022, from www.essentialchemi calindustry.org
[3]   Dunn Edwards Paints, 2013, What is Paints Made of? Retrieved February 10, 2022, from www.dunned wards.com.
[4]   Wikipedia, January 2022, Surfactants in Paint. Retrieved February 9, 2022, from https://en.wikipedia. org/
[5]   Ebbing and Gammon, 2015, General Chemistry 11th edition, Boston MA, Cengage.
[6]   https://coatings.specialchem.com/product/a-ashland-specialtychemical-natrosol-250-hhbr
[7]   National Library of Medicine, 2022, Sodium Bicarbonate. Retrieved March 9, 2022, from https://pubchem.ncbi.nlm.nih.gov.
[8]   Haz-Map, "Octoxynol, CAS# 9002-93-1", Haz-Map (2022, April 22), http://haz-map.com
[9]   https://chem.nlm.nih.gov/chemidplus/rn/9002-93-1
[10] JimTrade, "Octylphenol Polyglycol Ether Sulphate Sodium Salt", Jim Trade (2022, April 22), https://www.jimtrade.com/
[11] https://www.alfa.com/en/catalog/L15356/
[12] National Library of Medicine, 2022, Butyl Acrylate. Retrieved March 9, 2022, from https://pubchem.ncbi.nlm.nih.gov.
[13] Wikipedia, January 2022, Vinyl Acetate. Retrieved March 9, 2022, from https://en.wikipedia.org/
[14] National Library of Medicine, 2022, Potassium Persulfate. Retrieved March 9, 2022, from https://pubchem.ncbi.nlm.nih.gov.
[15] National Library of Medicine, 2022, Tert Butyl Hydroperoxide. Retrieved March 11, 2022, from https://pubchem.ncbi.nlm.nih.gov.
[16] Nandi U. S., Palit S. R., Hydrogen peroxide as initiator in vinyl polymerization in homogeneous system. I. Kinetic studies, Journal of Polymer Science, 1955, 17 (83), pp 65–78.
[17] National Library of Medicine, 2022, Sodium Formaldehyde Sulfoxylate. Retrieved March 12, 2022, fromhttps://pubchem.ncbi.nlm.nih.gov.
[18] https://coatings.specialchem.com/product/a-bruggemannreducing-agent-tp-1648
[19] Wikipedia, January 2022, Dibutyl Phthalate. Retrieved February 11, 2022, from https://en.wikipedia. org/
[20] Stefano Ciroi, 2017, Formaldehyde in Vinyl Adhesives. Retrieved March 13, 2022, from https://catas. com/en-GB/news.
[21] Michalak, K. Chojnacka, Biocides, Encyclopedia of Toxicology (Third Edition) 2014, Pages 461–463.

[22] National Library of Medicine, 2022, Vinyltrimethoxysilane. Retrieved March 9, 2022, from https://pubchem.ncbi.nlm.nih.gov.

[23] Momentive Solutions for a Sustainable Group, 2022, Silquest A-171 Silane. Retrieved March 12, 2022, from www.momentive.com.

[24] Encyclopedia Britannica, Inc., 2022, Emulsion Polymerization, Retrieved 10 March 2022, from www. Britannica.com

[25] Grechanovskii, V. A., Branching in Polymer Chains, *Rubber Chemistry and Technology* (1972) 45 (3): 519–545.

[26] *Charles E. C Jr., 2017*, Introduction to Polymers 4th edition, Boca Raton, Taylor & Francis Group.

[27] Swamy, R. N. and Tanikawa, S., 'An external surface coating to protect concrete and steel from aggressive environments', *Mater. Struct.* 26 (1993) 465–478.

[28] Tyssal, L. A., 'Industrial paints: Basic principles', (Pergamon Press Ltd, London 1964).

[29] Uemoto, K. L., Agopyan, V., Ranieri, R. and Quarcioni, V. A.,'Effect of concrete coating systems on chloride penetration', in 'Consec 98: Concrete Under Severe Conditions', Proceedings of an International Conference, Norway, (1998) 1321–1330.

[30] Uemoto KL, Agopyan V, Vittorino F. Concrete protection using acrylic latex paints: Effect of the pigment volume content on water permeability. Materials and Structures. 2001 Apr; 34(3):172–7.

3 Increase Product Quality for a Car-Wash Shampoo Concentrate

3.1 Overview

The market for vehicle cleaning products in Western Europe approached a value of $400 million dollars in 2007. Domestic and industrial automated cleaning of vehicles can include four main steps: prewash, main wash, rinse, and drying, besides the manual cleaning of domestic vehicles. This chapter presents the methods used to improve the properties of the current car wash shampoo concentrate formulation and increase its quality. The chapter also presents some new chemical compounds and additives that would increase the quality of the car wash shampoo concentrate and methods used to increase the concentration of the current formulation for its use and distribution.

3.2 Introduction

Saka International Group, a leading manufacturer of home care business, laundry, and car care products by Schnnell, ON, Canada, are looking to improve the chemical properties of their car wash shampoo. A liter of car wash shampoo concentrate dilutes only to 200 L using the current formulation. The formulation consists of the following chemical compounds: 5% linear alkyl benzene sulfonic acid (LABSA), 20% sodium lauryl ether sulfate (SLES), 5% sodium lauryl sulfate (SLS), 1% sodium hydroxide, 2% betaine, 61.5% water, 3% sodium triphosphate, 2% glycerol, and 0.5% propylene glycol.

Car maintenance products are classified into the interior and exterior car care products. Interior car care products are deodorants, grease cleaners, vinyl and plastic cleaners and polishes, and interior wins screen cleaners, carpet shampoos, and leather polishes. Interior car care products include tire dressings and cleaners, presoak detergents, car polishes, wash and wax formulations, windscreen cleaners, water-repellents and drying aids, and wheel rim cleaners. The climate and the season of the year affect the nature of the soiling of the vehicle and the ease of its removal.

The bodywork of the automobile consists of multiple coatings; each coating provides a variety of functions. Figure 3.1 shows the coating layers in the bodywork of modern cars. The paintwork of the vehicle is the external surface to be cleaned.

The base coat is usually water-based polymeric binders, fillers, and pigments. The inner coating, the electro-deposition paint, and the phosphate-based anticorrosion layer provide protection to the metal surface. On top of the protective coatings is the filler layer; it must have an excellent adhesion property to both the top coat and the base coat. The finishing lacquer must have good impact strength, retain gloss, and it must be waterproof. Domestic and industrial automated cleaning of vehicles can be

https://doi.org/10.1515/9783111316864-003

Base Lacquer
12–15μm

Filler Coating 35 μm

Electro-redeposition paint
18–25 μm

Phosphate anti-corrosion layer 2 μm

Metal bodywork

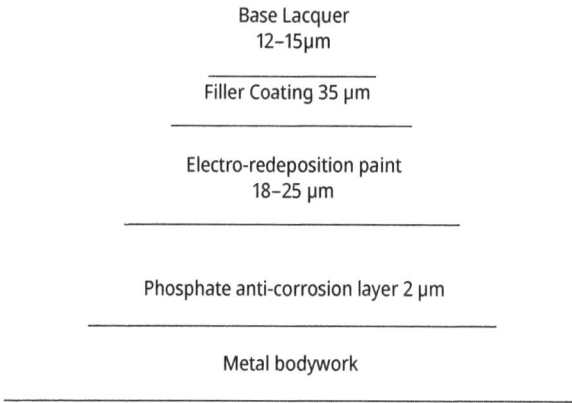

Fig. 3.1: Modern vehicle paintwork structure.

divided into four main steps: pre-wash, main wash, rinse, and drying, besides the manual cleaning of domestic vehicles. Prewash includes cold degreaser, microemulsion, and foam wash. The main wash includes shampoo and microemulsion. Rinse includes hot/cold wax and rinse aid.

Car shampoos can be either in liquid or powder form. Liquid car shampoos are a combination of binders, surfactants, and liquids dissolved in water as the main solvent. These products are easy to rinse off, high foaming, biodegradable, made to cut through grease on the bodywork, and they don't damage any part of the vehicle surface including the paintwork. Economy car shampoos do not contain builders. Powder car shampoos are made of a mixture of builders (carbonates, phosphates, or metasilicates) and surfactants (fatty alcohol ethoxylates or dodecylbenzene sulfonates) absorbed into the powder.

The main anionic detergent can be either alkylbenzene sulfonates and/or sodium lauryl ether sulfates. Sodium lauryl ether sulfate is incorporated into the formulation used when the denser, richer foam is required. Fatty acid alkanolamides, amine oxide, or betaine is used in the formulation for viscosity and to stabilize the foam produced. For greater foam stability and viscosity, amides are added to the formulation. To increase the quantity of the foam produced, betaines and amine oxides are used. Glycerol ether is used to ease grease removal. Secondary surfactants are used for viscosity and foam modifications and are also used to enhance spot removal and improve detergency. Binders such as phosphates (0.5–2.5%) are added to improve detergency.

Low hydrophilic-lipophilic balance (HLB) fatty alcohol ethoxylate/hydrotropic system replaced the traditional anionic surfactant-based car shampoo as they afford more effective cleaning performance, decreasing and they have low foam profile.

Sodium citrate is known to be a water softener and a PH adjuster. It is an ingredient used in most common liquid detergents. It is also used in some food products to adjust their acidity. It is used in ice cream, gelatin desserts, candy, and jelly. It is also

Tab. 3.1: Traditional anionic car wash shampoo formulation.

Chemical Compound	% by weight used the formulation
Soudium carbonate (binders)	2
Sodium metasilicate pentahydrate (binders)	3
Sodium citrate (water softner)	2
Glycerol ether (solvent)	4
Linear alkyl benzene sulphonate (30%) detergent	27
Sodium lauryl ether sulphate (28%) Detergent	10
Coconut diethanolamide (foam producer)	3
Water	49
Preservatives/ dyes	Q.S.

Tab. 3.2: HLB fatty alcohol ethoxylate/hydrotropic car wash shampoo formulation.

Chemical Compound	% by weight used the formulation
Fatty alcohol ethoxylate (low HLB)	5
Hydrophobe (alkyl glycoside or quaternary fatty amine ethoxylate)	5-10
TKPP (Tetrapotassium Pyrophosphate) Detergent builder	6
Sodium metasilicate	4
Balance water	

used in some pharmaceutical and personal care products such as sunscreens, facial moisturizers, makeup, baby wipes, soaps, shampoos, and conditioners.

Coconut diethanolamide is used as an emulsifying and foaming agent in personal care products and in cosmetics. It is extracted from coconut oil. It is also used in hydraulic fluids and industrial cooling lubricants.

3.3 Experimental

The chemical composition of the ingredients used in the making of the car-wash shampoo concentrate is presented in Tab. 3.1.

Linear alkyl benzene sulfonic acid is used as a mercerizing and washing agent in the textile industry. It is used as an emulsifier and wetting agent in small quantities with surfactants as it increases the surface area of distempers. Because of its good performance and low-cost linear alkylbenzene sulfonic acid is the largest volume synthetic anionic detergent. As all surfactants, linear alkylbenzene sulfonic acid has both hydrophilic and hydrophobic groups. Other examples of commercial anionic surfactants are alkyl sulfates and alpha-olefin sulfonates. These compounds are produced by sulfonation, and they are nonvolatile compounds. Linear alkylbenzene sulfonic acid consists of a phenyl isomer of five to two-position substituents, different alkyl chain lengths consisting of 10–14 carbon atoms (C10–C14), and an aromatic ring sulfonated at the para-position attached to the linear alkyl chain at any position except position 1, the 1-phenyl position. The chemical and physical properties of linear alkylbenzene sulfonic acid differ based on the length of the alkyl chain, giving rise to different formulations and different usage in various applications.

Sodium lauryl ether sulfate is a surfactant, and it is an anionic detergent. It is a very effective foaming agent, inexpensive, and used in shampoos, toothpastes, and soaps.

Sodium lauryl sulfate is used in hygiene, cleaning, food, and pharmaceutical products, and it is an anionic-surfactant. It is widely used as a food additive in the food industry and as an emulsifier and an ionic solubilizer in the pharmaceutical industry.

Sodium hydroxide sulfate is a strong base, hygroscopic solid, and soluble in water and can cause severe burns. It is used in various industries; it is used in the manufacture of drain cleaners, soaps, detergents, textiles, drinking water, pulp, and paper. It is used as a paint stripper, cleaning agent, relaxer, and in food preparation.

Betaine is a chemical compound occurring in plants, and it is an amino acid; it is a white solid at room temperature. It is present in living cells, and it is a methylated nitrogen compound. It is used in the pharmaceutical industry in the preparation of shampoo and soap as it is a nonionic surfactant. Due to its high surface activity, its derivatives are used as efficient cleansing agents. It is also used as viscosity modifiers, foam stabilizers, and detoxifiers. Betaine esters can be used in antiperspirants as it processes antimicrobial activity.

Water is an inorganic compound that is odorless, tasteless, transparent liquid at room temperature and acts as a solvent. It is present in 70% of the earth's surface as seas and oceans. In the world economy, 70% of water is used in agriculture.

Sodium triphosphate is used as a component of industrial and domestic products on a large scale.

It is used as a builder and a water softener in commercial detergents. Detergents are deactivated in water containing high concentrations of Mg^{2+} and Ca^{2+}, hard water. It is a chelating agent as it binds tightly to bications and prevents them from interfering with sulfonate detergent. It is used as an emulsifier to retain moisture, and it is also used as a preservative in the food industry. It is also used as an anticracking agent, flame-retardant, anticorrosion pigment, synthetic tanning agent, and masking agent in the leather industry.

Glycerol is an odorless and nontoxic colorless viscous liquid at room temperature. Glycerol is hygroscopic in nature and is miscible in water. It is used as a humectant in pharmaceutical formulations as it improves the ability of the skin to absorb water, and it is used in the food industry as a sweetener.

Propylene glycol is a colorless viscous liquid at room temperature and has a faintly sweet taste. It is miscible in a wide variety of solvents including chloroform, acetone, and water. It is a nonirritating substance with low volatility. It is used in various industries, including food and drug, antifreezes, polymers, and electronic cigarettes. It is used as a humectant in hand sanitizers to prevent skin drying. It is used in coffee-based drinks, ice creams, whipped dairy products, soda, and liquid sweeteners. Polypropylene glycol alginate gives rise to a greater increase in foam stability equal to the amount of neutral polysaccharides.

The current formulation consists of 1% builder in the form of sodium hydroxide, 3% water softener in the form of sodium triphosphate, 32% surfactants in the form of 5% linear alkyl benzene sulfonic acid, 20% sodium lauryl ether sulfate, and 5% sodium lauryl sulfate and 2.5% solvents in the form of 0.5% propylene glycol and 2% glycerol.

Sodium and potassium hydroxides are known to be used in wheel rim cleaner formulations as the alloy wheel picks up dirt and grease from the road, and they are prone to dirt from the abrasive wear of brake shoes. The amount of sodium and potassium hydroxides used in the wheel rim cleaner formulation is 0–15%. It isn't known to be used in car wash formulations. Other builders are known to be used in car wash formulations to soothe out slight imperfections and to remove road grease and stubborn tar from the bodywork of the vehicle. These builders would be calcium carbonate, silicones, and lamella aluminum silicates. To increase the quality of the car wash shampoo formulation, silicone derivative builders are added to the formulation. Silicone derivative builders contribute to the ease of application of the products, the gloss, and its water-repellency property.

The amount of surfactant used in the current car wash shampoo formulation concentrate is 32% (5% linear alkyl benzene sulfonic acid, 20% sodium lauryl ether sulfate, 5% sodium lauryl sulfate, and 2% betaine). To increase the quality of the car wash shampoo, the amount of surfactants in the formulation must be increased to 37%.

The solvents used in the current formulation: 0.5% propylene glycol and 2.0% glycerol are both known to be used as humectants in pharmaceutical and personal care product formulations to improve the ability of the skin to absorb water. These solvents aren't known to be used in car wash shampoo formulations. Other solvents are used in car wash shampoo formulations (4% glycol ether) to aid in the removal of surface dirt and to act as a carrier for silicates and components of the shampoo. Solvents that are commonly used in car wash shampoo formulations are glycol ethers such as dipropylene alcohol monomethyl ether to dissolve grease.

The choice of solvent is critical to avoid the stress and cracking of plastics and to avoid damage to the painted surface. It didn't recommend the use glycerol or propylene glycol in the formulation and glycol ether. Also, the amount of solvents used in the current formulation is low, 2.5%, compared to the amount used, reference 4%. The amount of solvents used should be increased to 4% to increase the quality of the car wash shampoo formulation.

For the current formulation to dilute to a greater volume of 500 L and to also maintain the quality of the car wash shampoo concentrate, the amount of surfactants,

Tab. 3.3: Chemical name, molecular formula, and molecular weight of chemical compounds used in the current formulation.

Chemical Name	Chemical Structure	Molecular Formula	Molecular weight (g/mol)	Amount Used in the Current Formulation (%)
Linear Alkyl Benzene Sulphonic Acid (LABSA)		$CH_3(CH_2)_{11}C_6H_4SO_3H$	326.49	5
Sodium Lauryl Ether Sulphate (SLES)		$CH_3(CH_2)_{11}(OCH_2CH_2)_nOSO_3Na$	Variable; typicaly around 421 g/mol	20
Sodium Lauryl Sulphate (SLS)		$C_{12}H_{25}NaSO_4$	288.372	5
Sodium Hydroxide		$NaOH$	39.997	1
Betaine		$C_5H_{11}No_2$	117.146	2
Water (Dihydrogen Monoxide)		H_2O	18.01	61.5

| Sodium triphosphate | | $Na_5P_3O_{10}$ | 367.864 | 3 |

| Glycerol | | $C_3H_8O_3$ | 92.094 | 2 |

| Propylene Glycol | | $C_3H_8O_2$ | 76.095 | 0.5 |

builders, foaming agents, and solvents must be increased relative to the amount of the main solvent used to dissolve its components. The current formulation dilutes to 200 L; in order for the formulation to dilute to double this volume, the amount of the main solvent used to dissolve its components must be decreased to half, and the amount of surfactants/builders/foaming agents and solvents must be increased by 1½. Taking the traditional anionic car wash shampoo formulation as an example, the percentage of each chemical compound in the formulations would be 3% sodium carbonate, 4.5% sodium metasilicate pentahydrate, 3% sodium citrate, 6% glycerol ether, 40.0% linear alkyl benzene sulfonate, 15% sodium lauryl ether sulfate, 4.5% coconut diethanolamide, and 24.0% water as the dissolving solvent.

3.4 Conclusion

To increase the quality of the car wash shampoo-concentrate 2% builders in the form of sodium carbonate, 3% builders in the form of silicates, and 3% foaming and emulsifying agents in the form of coconut diethanolamide should be included in the formulation. It also recommends the use of another different solvent to dissolve grease and improve the quality of the car wash shampoo concentrate, 4% glycol ether.

Foaming and emulsifying agent are used to dissolve solvents, builders, and silicates and facilitate the formation of foams. Builders used in car wash shampoo formulations remove road grease and stubborn tar from the bodywork of the vehicle, smooth out minor surface scratches and slight imperfections, contribute to the gloss, the water repellency, increase the durability of the overseal, and polish the paintwork. The choice of builders and solvents is critical to avoid the stress-cracking of plastics and to avoid damage to the painted surface. Calcium carbonates, silicones, lamella aluminum silicates, and poly dimethyl siloxane are known to be used in car wash shampoo formulations as builders.

It recommends avoiding the use of sodium hydroxide as a builder or the use of glycerol/propylene glycol as solvents in the car wash shampoo formulation, and the amount of surfactants must be increased to at least 37%.

Sodium hydroxide is known to be used in wheel rim cleaner formulation as the alloy wheel pick up dirt and grease from the road not in car wash shampoos; glycerol and propylene glycol are known to be used as humectants in lotions and personal care products. Sodium hydroxide, glycerol, and propylene glycol aren't known to be used in car wash shampoo formulations.

In order to increase the concentration of the car-wash shampoo concentrate and make a liter of concentrate dilute to 500 L instead of 200 L, the amount of builders, solvents, and surfactants must be increased to 1½, and the amount of water must be decreased to half.

Chapter Questions

1) What's the aim of the chapter?
2) What are the types of car wash shampoos and the main difference in their chemical composition?
3) What's the main chemical composition of an anionic car was shampoos?
4) What are the chemicals used in the current formulation? What are the methods used to increase its quality? And what are the recommended additives to increase its quality?
5) What's the chemical structure and the main function of glycerol in a car wash shampoo?

References

[1]   http://www.sakagroup.com
[2]   Company M., Karsa D. R., Handbook for Cleaning/ Decontamination of Surfaces, Amsterdam, the Netherlands, Elsevier B.V., 2007.
[3]   H.G. Hauthal and G. Wagner (eds.), Household Cleaning, Care and Maintenance Products: Chemistry, Application, Ecology and Consumer Safety, pub1. Verlag Fur chemische Industrie H Ziolkowsky GmbH, 2004.
[4]   Surfactants Selector; A Guide to the Selection of I&I and Household Product Formulations, Akcros Chemicals (now part of Akzo Nobel Surface Chemistry AB) 1998.
[5]   Further formulation information available from Akzo Nobel Surface Chemistry AB, S 444 85 Stenungsund, Sweden, on request.
[6]   Chemicalland21, 2013, Linear Alkylbenzene Sulfonic Acid, Retrieved May 2022, from www.chemical land21.com/specialtychem/perchem.
[7]   Wikipedia the Free Encyclopedia, 2022, Sodium Laureth Sulfate, Retrieved May 2022, from https://en.wikipedia.org.
[8]   Wikipedia the Free Encyclopedia, 2022, Sodium Dodecyl Sulfate, Retrieved May 2022, from https://en.wikipedia.org.
[9]   Wikipedia the Free Encyclopedia, 2022, Sodium Hydroxide, Retrieved May 2022, from https://en.wikipedia.org.
[10] Wikipedia the Free Encyclopedia, 2022, Trimethylglycine, Retrieved May 2022, from https://en.wikipedia.org.
[11] Wikipedia the Free Encyclopedia, 2022, water, Retrieved May 2022, from https://en.wikipedia.org.
[12] Wikipedia the Free Encyclopedia, 2022, Sodium Triphosphate, Retrieved May 2022, from https://en.wikipedia.org.
[13] Wikipedia the Free Encyclopedia, 2022, Glycerol, Retrieved May 2022, from https://en.wikipedia.org.
[14] Wikipedia the Free Encyclopedia, 2022, Propylene Glycol, Retrieved May 2022, from https://en.wikipedia.org.
[15] World of Chemicals, 2022, Linear Alkylbenzene Sulfonic Acid, Retrieved May 2022, from www.worldofchemicals.com.
[16] Valappil K., Lalitha S., Gottumukkala D., Sukumaran R. K., Pandey A., White Biotechnology in Cosmetics, Industrial Biorefineries & White Biotechnology, 2015, pp. 607–652. ISBN 9780444634535

[17] "Water Q&A: Why is water the "universal solvent"?". http://www.usgs.gov. (U.S. Department of the Interior). Retrieved 15 January 2021.

[18] "CIA – THE WORLD FACTBOOK Geography Geographic overview". Central Intelligence Agency. Retrieved 20 December 2008.

[19] Baroni, L.; Cenci, L.; Tettamanti, M.; Berati, M. (2007). "Evaluating the environmental impact of various dietary patterns combined with different food production systems". European Journal of Clinical Nutrition. 61 (2): 279–286. doi:10.1038/sj.ejcn.1602522. PMID 17035955.

[20] Complexing agents, Environmental and Health Assessment of Substances in Household Detergents and Cosmetic Detergent Products, Danish Environmental Protection Agency, Accessed 2008-07-15

[21] Schrödter, Klaus; Bettermann, Gerhard; Staffel, Thomas; Wahl, Friedrich; Klein, Thomas; Hofmann, Thomas (2008). "Phosphoric Acid and Phosphates". Ullmann's Encyclopedia of Industrial Chemistry. doi:10.1002/14356007.a19_465.pub3. ISBN 978-3527306732.

[22] Oxford Dictionaries – English, glycerol – Definition of glycerol in English by Oxford Dictionaries". Archived from the original on 21 June 2016. Retrieved 21 February 2022.

[23] Christoph, Ralf; Schmidt, Bernd; Steinberner, Udo; Dilla, Wolfgang; Karinen, Reetta (2006). "Glycerol". Ullmann's Encyclopedia of Industrial Chemistry. doi:10.1002/14356007.a12_477.pub2. ISBN 3527306730.

[24] Sullivan, Carl J.; Kuenz, Anja; Vorlop, Klaus-Dieter (2018). "Propanediols". Ullmann's Encyclopedia of Industrial Chemistry. Weinheim: Wiley-VCH. doi:10.1002/14356007.a22_163.pub2.

[25] Lohrey, Jackie. "Ingredients in Hand Sanitizer". LIVESTRONG.COM. Retrieved 2018-06-11.

[26] "Quackmail: Why You Shouldn't Fall For The Internet's Newest Fool, The Food Babe". Butterworth, Trevor. Forbes. 16 June 2014. Retrieved 18 March 2015.

[27] G. Jackson, R. T. Roberts and T. Wainwright (January 1980). "Mechanism of Beer Foam Stabilization by Propylene Glycol Alginate". Journal of the Institute of Brewing. 86 (1): 34–37. doi:10.1002/j.2050-0416.1980.tb03953.x.

[28] PersianUTab, "Sodium citrate in detergents", www.persianutab.com, 2020. Retrieved May 11, 2022

[29] Contact Dermatitis Institute, "Coconut diethanolamide", www.contactdermatitisinstitute.com, 2022. Retrieved May 11, 2022.

4 Modification to an Acrylic Paint Formulation

4.1 Overview

Paints are used to protect and prolong the life of natural and synthetic materials as it acts as a barrier against environmental conditions. Paints contain extenders, solvents, pigments, binders, and some additives. The contents of acrylic white matt emulsion paint are known to be 25% pigments, 12% extender pigments, 5% additives, 44% solvents, and 14% binders.

Binders are matrices and are used to hold the pigment in place. Extenders have larger pigment particles to improve adhesion, strengthen the film, and save the binder. Pigments are used to give color and opacity to the paint. A solvent can be either an organic solvent or water and is used as a thinner to dissolve paint components and make them uniform. Additives are commonly used to improve the properties of the paint.

This chapter briefly shows how to manipulate the chemical and physical properties of chemical compounds making acrylic matt paint improve its quality. It presents the chemical compounds that would potentially lower the scrub resistance test. It also presents the additives and the chemical compounds that would potentially increase the quality and increase the scrub resistance test for acrylic matt paint.

4.2 Introduction

A polymer production unit in Saudi Arabia, PSCF, is looking to improve its current formulation of the current acrylic matt paint formulation as the current formulation shows a low scrub resistance test compared to the reference. The current paint scrub-resistance test shows that the paint operates with 60% effectiveness compared to the reference.

The current formulation consisted of 47% vinyl acetate, 0.1258% potassium persulfate, 0.0234% tertiary butyl hydrogen peroxide, 0.06% hydrogen peroxide, 0.05% sodium formaldehyde sulfoxylate, 1.0% di butyl phthalate, 0.1% formaline, 0.05% biocide, 0.1% Silquest A-171 and 0.06% defoamer, 0.4% Natrosol 250-HHBR, 44.39% water, 0.14% sodium bicarbonate, 2.0% polyoxyethylene 25 octyl phenol, 2.0% octyl phenol polyglycol ether sulfate sodium salt, 0.5%, Provichem, and 2.0% butyl acrylate. Chemical names, molecular formulas, and molecular weights of all chemical compounds used in making the current acrylic paint formulation are listed in Tab. 4.1.

An acrylic white matt emulsion paint is known to consist of 25% pigments, 44% solvents, 12% extender pigments, 5% additives, and 14% binders (Fig. 4.1).

https://doi.org/10.1515/9783111316864-004

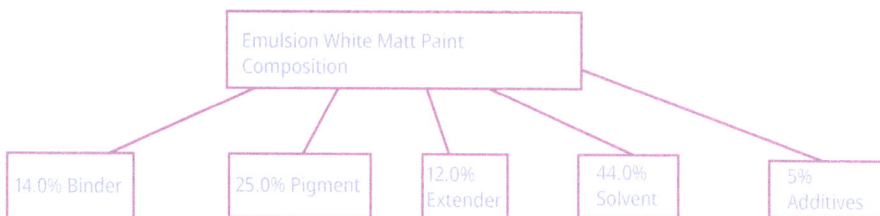

Fig. 4.1: Chemical composition for a matt white paint.

Tab. 4.1: Chemicals, amounts in percentage, and the role of chemical compounds used in the current acrylic paint formulation.

Chemical Name	Percent (%) Used in the Current Formulation	Molecular Formula	Role of Each Chemical Compound in Current Formulation
Water	44.09	H2O	Solvent
Natrolsol 250-HHBR	0.4	$C_{36}H_{70}O_{19}$	Nonionic water thickener
Sodium Bicarbonate	0.14	$NaHCO_3$	Pigment
Polyoxyethylene (25) octyle phenyl ether	2.0	(C2-H4-O)mult-C14H22-O	Surfactant and buffer
Octyl phenol polyglycol ether sulfate sodium salt	2.0	Octylphenol polyglycol ether sulfate sodium salt	Surfactant and buffer
Provichem (*Sodium vinylsulfonate*)	0.5	$C_2H_3NaO_3S$	Surfactant
Buyl acrylate	2.0	$C_7H_{12}O_2$	Monomer in homo or copolymerizaton (Binder)
Vinyl acetate	47	$C_4H_6O_2$	Monomer in homo or copolymerization (Binder)
Potassium Persulfate	0.1258		Initiator
Tertiary butyl hgrogen peroxide	0.0234	$C_4H_{10}O_2$	Oxidizing catalyst
Hydrogen peroxide	0.06	H_2O_2	Initiator
Sodium formaldehyde sulfoxylate	0.05	CH_7NaO_5S	Reducing agent
DI butyl phthalate	1.0	$C_{16}H_{22}O_4$	Secondary plasticizer
Formaline	0.1	CH_2O	Preservative
Biocide	0.05	Biocide	Biocide

Tab. 4.1 (continued)

Chemical Name	Percent (%) Used in the Current Formulation	Molecular Formula	Role of Each Chemical Compound in Current Formulation
Silquesit A-171 (Vinyltrimethoxysilane)	0.1	$C_5H_{12}O_3Si$	Crosslinking polymer
Defoamer	0.06	Defoamer	Defoamer
Total Percentage	100	N/A	N/A

4.3 Detailed Analysis

The current acrylic paint formulation consisted mainly of 44.39% water as a solvent, 49% pre-emulsion monomers (vinyl acetate and butyl acrylate), and 6.61% additives (Natrosol 250-HHBR, sodium bicarbonate, polyoxyethylene 25 octyl phenol, octyl phenol polyglycol ether sulfate sodium salt, Provichem, 2.0% butyl acrylate, potassium persulfate, tertiary butyl hydrogen peroxide, hydrogen peroxide, sodium formaldehyde sulfoxylate, dibutyl phthalate, formaline, biocide, Silquest A-171, and defoamer).

4.4 Conclusion

The current formulation consists of pigments 0.14% in the form of sodium bicarbonate compared to the literature amount 25%. Extender pigments weren't used in the current formulation compared to the literature amount 12%, surfactants 4.5% compared to the literature amount 0–1%, and binders 49% compared to the literature amount 14%. The copious amounts surfactants are causing a low-scrub resistance test of the current formulation compared to the reference.

In order to increase the quality of the acrylic paint, pigments, extenders, and another type of pre-emulsion monomer should be included in the formulation to form a firm protective layer and prevent the dissolution of the paint film from the surface and improve the scrub resistance test of the acrylic paint. It is recommended to avoid using copious amounts of surfactants, 4.5% (2.0% octyl phenol polyglycol ether sulfate sodium salt, 2.0% octyl phenol polyglycol ether sulfate sodium salt, and 0.5% Provichem).

Chapter Questions

1) What are uses for paints?
2) What's the role of sodium bicarbonate in the current formulation?
3) What's the chemical composition for white paint?
4) What are the chemical compounds found to be missing from the current formulation decreasing its quality?
5) What are the chemical compounds mainly found to be causing the low scrub resistance in the current formulation?

References

[1] http://pscf.com.sa/
[2] Izzo, Francesca Caterina; Balliana, Eleonora; Pinton, Federica; Zendri, Elisabetta, "A preliminary study of the composition of commercial oil, acrylic and vinyl paints and their behavior after accelerated ageing conditions", 2014, 353–369. doi:10.6092/ISSN.1973-9494/4753.
[3] Abdel-Wahab H., Gund, M. Chemical Formulations for Acrylic Matt and Acrylic Gloss Paints. American Journal of Applied and Industrial Chemistry. Vol. 6, No. 1, 2022, pp. 13–19. doi: 10.11648/j.ajaic.20220601.13
[4] Nandi U. S., Palit S. R., Hydrogen peroxide as initiator in vinyl polymerization in homogeneous system. I. Kinetic studies, Journal of Polymer Science, 1955, 17 (83), pp 65–78.
[5] I. Michalak, K. Chojnacka, Biocides, Encyclopedia of Toxicology (Third Edition) 2014, Pages 461–463.

5 Modification to a Chemical Formulation for Optimal Performance

5.1 Overview

In a chemical reaction, the reactant substance changes to a product substance, and the product substance will have different physical and chemical properties than the reactant substance. All chemical reactions involve a detectable change: color change, bubbling, heat evolution, heat absorption, light emission, or formation of a precipitate. Reactions are classified into three main reactions: precipitation reactions, acid-base reactions, and oxidation-reduction reactions. The counting unit for a number of atoms, ions, or molecules in a laboratory seize sample is mole, abbreviated mol. A limiting reactant is a reactant that is completely consumed in a chemical reaction; it limits and determines the amount of product formed; the other reactants are called excess reactants. The quantities of product formed and reactant consumed are restricted by the amount of limiting reactant. This chapter presents the methods used to increase the quality of a bubble-forming product, increasing the amount of bubbling per table by using the difference in chemical and physical properties of the chemical compounds included in the ingredient. The chapter also presents chemical compounds in the ingredient that would be harmful to the quality of the product, and it also presents new additives that would increase the quality of the product.

5.2 Introduction

Some of the companies in the US are looking to produce kid-safe foaming tablets from household chemicals that would produce bubbles when placed in the snow. The ingredient included the following chemical compounds: corn starch (pure starch), red beet powder (betanin), salt (sodium chloride), cream tartar (potassium bitartrate), baking soda (sodium bicarbonate), vitamin C (citric acid), and drops of water (H_2O). The amount of bubbling was minimal; also, the tablet didn't hold together and was grainy and fragile. Since bubbling forms when the fragile tablet is placed in the snow, it is the evidence of a chemical reaction.

In a chemical reaction, the reactant substance changes to a product substance, and the product substance will have different physical and chemical properties than the reactant substance. All chemical reactions involve a detectable change.

A chemical equation is used to describe a chemical reaction. A chemical reaction is represented by a molecular equation. Reactant species are always listed on the left side of a chemical equation and the product species are listed on the right side of the chemical equation and they are separated by an arrow. A chemical reaction can also be presented by an ionic equation or net ionic equation:

https://doi.org/10.1515/9783111316864-005

Reactants $\rightarrow$ Products

The reactions are classified into three main reactions: precipitation reactions, acid-base reactions, and oxidation-reduction reactions. Precipitation reactions would involve the formation of insoluble ionic compound forms from mixing a solution of two ionic substances. Acid-base reactions involve the reaction of an acid with a base and a transfer of a proton between reactants is involved. Oxidation-reduction reactions (redox reactions) involve the transfer of an electron between reactant species.

The reaction that takes place when placing the created tablet in water/ice is an acid-base reaction. Acid-base reactions are widely known for their ability to produce bubbling if either reactant is a weak acid or weak base that would produce a gas product. In an acid-base reaction, acid and base properties neutralize, where the anion of the acid and the cation of the base combine to form the salt, and the hydroxide anion, OH^-, combines with the hydrogen cation, H^+, to form water.

The counting unit for a number of atoms, ions, or molecules in a laboratory seize sample is mole, abbreviated mol. A mole of samples of different substances has different masses. For example, 1 mol of 24 mg and 1 mol of 12C. A single 24 mg atom is twice as massive and has a mass of 24 a.m.u, as a single 12C atom has a mass of 12 a.m.u. The formula weight of any substance is numerically equal to the molar mass of that substance. For example, for table salt with the chemical formula NaCl, the formula weight is 58.5 a.m.u, and its molar mass is 58.5 g/mol. The molar mass of a substance in gram per mole can be used as a conversion factor to convert the mass of a substance into moles or moles of a substance into grams.

The relative number of molecules in a reaction is represented by coefficients in a balanced chemical equation. The relative number of moles and the relative number of molecules in a reaction are indicated by the coefficients in a balanced chemical equation.

A limiting reactant is a reactant that is completely consumed in a chemical reaction. It limits and determines the amount of product formed. The other reactants are called excess reactants. The quantities of product formed and reactant consumed are restricted by the amount of limiting reactant. If one reactant, the limiting reactant, is consumed, the reaction stops, and the amount of limiting reactant becomes zero at the end of the reaction.

Carboxylic acids are the largest category of weak organic acids that contain carboxylic groups (–COOH). The generic formula for carboxylic acids is R-COOH, where R = alkyl group.

The acidic behavior of carboxylic acids is due to the oxygen atom bonded to the carboxyl group carbon drawing electron density from the hydroxyl bond. It helps stabilize the conjugate base and increases the polarity of the –OH bond. The carboxylate anion, the conjugate base of acid, has resonance, which spreads the negative charge over several atoms and contributes to the stability of the anion.

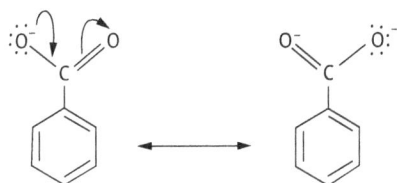

Fig. 5.1: Resonance behavior of carboxylic acids.

Alcohols are compounds that contain one or more hydroxyl or alcohol groups (−OH). The −OH bonds are polar; bonds and alcohols are more soluble in polar solvents than hydrocarbons, and OH groups can form hydrogen bonds: Examples of alcohols are 1,2

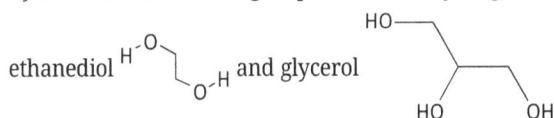

ethanediol and glycerol

In industry, occasional modification to the formulation of a product is necessary to improve its quality and performance. In 2021 Chegeni et al. [18] developed a new formulation for air revitalization tablets using the Taguchi statistical method in closed atmospheres. The air revitalization tablets are used in closed spaces such as submarines, underground mines, shelters, and spacecrafts to absorb carbon dioxide, decrease its concentration, and produce fresh oxygen necessary for life in a closed space. The initial formulation consisted of various amounts of additives: MnO_2 (catalyst), $LiOH$ (CO_2 auxiliary absorbent), $Cu_2Cl\,(OH)_3$, and $CaSO_4$. It has been found that for optimal performance, the greatest amount of oxygen is produced, and the maximum amount of carbon dioxide adsorbed occurs when using a certain amount of each chemical compound in the formulation. This optimal formulation consisted of 3 wt% lithium hydroxide, 3 wt% copper oxychloride, 3 wt% calcium sulfate, and 5 wt% manganese dioxides.

In medicine, modification to the method of delivery of the drug to the organ is necessary to improve its performance. In 2020, Chen et al. [19] developed an optimal method to deliver the drug formulation to the pancreas; they used a bubble bursting-mediated oral drug delivery system that enables concurrent delivery of hydrophilic and lipophilic chemotherapeutics for treating pancreatic tumors in rats. The optimal formulation consisted of paclitaxel (PTX; 1 mg) that had been predissolved in capric acid (18 mg), gemcitabine (GEM; 1 mg), sodium bicarbonate (5 mg), and citric acid (0, 2, 4, 6, or 8 mg) was exposed to a simulated intestinal fluid (SIF; 3 mL, pH 6.4) that contained bile extract (5 mg/mL). The oral drug delivery system that initiates an effervescent reaction to form gas-bubble carriers was proposed; these carriers deliver lipophilic PTX and hydrophilic GEM in the small intestine. The bursting of the bubbles promotes the absorption of the drugs in the intestines. The study revealed that the orally delivered formulation has no toxic side effects that are associated with the i.v.-injected formulation. It resulted in an increase in the bioavailability of PTX, and the oral formulation had a greater impact than the i.v. formulation on tumor-specific stromal depletion, resulting in greater inhibition of tumor growth with no evidence of metastatic spread. This unique approach of oral chemotherapy has the potential for use on outpatients, greatly improving the quality of their lives and enhancing therapeutic efficacy.

5.3 Experimental

The chemical composition of the ingredients used in the making of the bubble-producing tablets is presented in Tab. 5.1.

Corn starch is pure starch and is known to be used as a liquid thickener. Red beet powder is used as a dye that colors the tablet. Table salt is sodium chloride with the formula NaCl and molar mass of and this salt has a neutral pH, pH = 7. Cream tartar is potassium bitartrate with formula $KC_4H_5O_6$ and molar mass: 188.177 g/mol. Baking soda is a weak base with a formula of $NaHCO_3$ and a molar mass: 84.007 g/mol. Citric acid is a triprotic acid, a weak organic acid that has the formula of $C_6H_8O_7$ and a molar mass: 192.124 g/mol.

Tab. 5.1: Chemical name, molecular formula, and molecular weight of household chemicals in the ingredients.

Household Chemical and chemical Name	Chemical Structure	Molecular Formula	Molecular Weight (g/mol)
Corn Starch (Pure Starch)		$C_{21}H_{48}O_{20}$	692.66
Table Salt (Sodium Chloride)		NaCl	58.44

Tab. 5.1 (continued)

Household Chemical and chemical Name	Chemical Structure	Molecular Formula	Molecular Weight (g/mol)
Cream Tartar (Potassium Bitartrate)		$KC_4H_5O_6$	188.177
Baking Soda (Sodium Bicarbonate)		$NaHCO_3$	84.007
Citric Acid (Vitamin C)		$C_6H_8O_7$	192.124
Water (Dihydrogen Monoxide		H_2O	18.01
Red Beet (Betanin)		$C_{24}H_{26}N_2O_{13}$	550.47
Glycerol		$C_3H_8O_3$	92.094

The bubbling while creating the tablet using water as a binding agent is due to the reaction between weak acid (citric acid) and potassium bitartrate with the weak base (baking soda) to form a salt, water, and carbon dioxide, causing loss of bubbles while making the tablet.

In order to get the maximum amount of yield, the maximum amount of CO_2 per tablet, intern the greatest amount of bubbling, the number of moles of citric acid to sodium hydrogen carbonate must be 1:3, and the number of moles of sodium bitartrate to sodium hydrogen carbonate must be 1:1 according to the balanced chemical equations.

Vitamin C, citric acid, is a triprotic acid containing three acidic groups and one hydroxyl group. The hydroxyl groups (−OH) are alcoholic groups, and they don't react with sodium hydrogen carbonate or baking soda, while the acidic groups, carboxylic group (−COOH), can react with sodium hydrogen carbonate to form a salt, carbon dioxide, and water. The bubbling is due to the formation of carbon dioxide as a product.

The molar amount of citric acid and sodium bicarbonate reactants must be 1:3 in order to produce the maximum amount of bubbling in the formula per tablet:

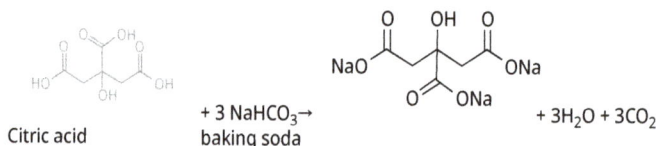

Citric acid   + 3 NaHCO$_3$→
               baking soda

Fig. 5.2: Acid-base reaction between citric acid and backing soda.

Using 1 g of vitamin C with formula $C_6H_8O_7$ in the ingredients, the amount needed for a reaction to go to completion and produce the maximum amount of carbon dioxide, and hence the maximum amount of bubbling would be 1.3 g of baking soda as follows:

$$1\,g\,/192.124\,g/mol \times 3\,NaHCO_3\,/1\,C_6H_8O_7 \times 84.007\,g/mol = 1.312\,g\,of\,NaHCO_3$$

Cream tartar, potassium bitartrate, is a monoprotic acid containing an acidic group and two hydroxyl groups. The hydroxyl groups (−OH) are alcoholic groups, and they don't react with sodium hydrogen carbonate or baking soda, while the acidic group, carboxylic group (−COOH), can react with sodium hydrogen carbonate to form salt, carbon dioxide, and water. The bubbling is due to the formation of carbon dioxide as a product.

Using 1 g of cream tartar, potassium bitartrate with formula $KC_4H_5O_6$ in the ingredients, the amount needed for a reaction to go to completion and produce the maximum amount of carbon dioxide, and hence the maximum amount of bubbling would be grams of baking soda as follows:

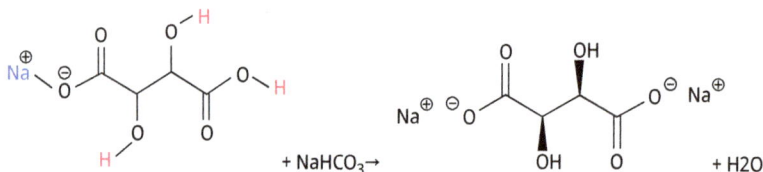

+ NaHCO$_3$→   + H2O

Fig. 5.3: Acid-base reaction between cream tartar and baking soda.

$1\,\mathrm{g}/188.17\,\mathrm{g/mol} \times 1\,\mathrm{NaHCO3}/1\,\mathrm{KC_4H_5O_6} \times 84.007\,\mathrm{g/mol} = 0.4464\,\mathrm{g}$ of $\mathrm{NaHCO_3}$.

The number of grams of sodium hydrogen carbonate needed to completely neutralize 1 g of vitamin C and 1 g of sodium bitartrate must be 1.3117 g +0.4464 g = 1.76 g.

Water as a binding agent causes bubble loss while creating the tablet, and glycerol would act as a better binding agent since sugars would cause bubbling to persist, and the compounds in the ingredients are inert in glycerol. Sodium chloride reduces bubble formation, but the amount of base used was insufficient to neutralize both citric acid and potassium bitartrate in the formula.

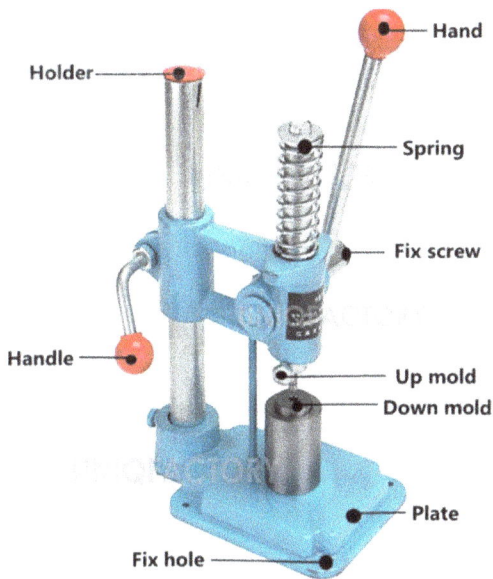

Fig. 5.4: Manual tablet press machine.

5.4 Conclusion

For optimal performance, the molar amounts of reactants were calculated to produce the greatest amount of bubbling per tablet. The bubbling power of the tablets in water is due to the reaction between vitamin C and cream tartar with baking soda. Using 1 g of vitamin C and 1 g of sodium bitartrate, 1.76 g of sodium hydrogen carbonate is required to be used as an ingredient to get the maximum amount of yield, the maximum amount of CO_2 in the product, in turn, the maximum amount of bubbling per tablet in the formula.

Sugars can be used to increase the quality of the product. It is recommended to use a greater number of moles of starch for the tablet to be thicker on snow, and glycerol acts as an ideal chemical binding agent for the ingredient.

Sugars, including glycerol, sugar alcohol, and sucrose, are good additives to cause the persistence of the bubbling forming ability of the tablet created, and it is not recommended to use sodium chloride in the formulation as it would prevent forming bubbles.

For the tablet to look smoother and for equal distribution of each of the compounds in the ingredient per table, the ingredients must be grinded to powder before adding the binding agent glycerol.

The tablets have been created using the correct molar amount for each chemical in the ingredients, and the amount of bubbling has increased significantly; the number of bubbles formed has doubled using the correct formula.

Chapter Questions

1) What's the aim of the chapter?
2) What's the chemical name for cream tartar?
3) What are the chemicals used in the current formulation?
4) What's the type of reaction occurring when the tablet is placed in aqueous medium?
5) What's the molar amount in grams of baking soda required for the optimal performance of the current formulation, 1g of citric acid and a gram of cream tartar?

References

[1]   https://en.wikipedia.org/wiki/Acid%E2%80%93base_reaction
[2]   https://www.molinstincts.com/formula/Cornstarch-cfml-CT1087471098.html
[3]   https://en.wikipedia.org/wiki/Sodium_chloride
[4]   https://en.wikipedia.org/wiki/Potassium_bitartrate
[5]   https://en.wikipedia.org/wiki/Sodium_bicarbonate
[6]   https://pubchem.ncbi.nlm.nih.gov/compound/Citric-acid
[7]   https://en.wikipedia.org/wiki/Betanin
[8]   Ebbing and Gammon, 2015, General Chemistry, Cengage, P.103
[9]   Ebbing and Gammon, 2015, General Chemistry, Cengage, P.108
[10] Ebbing and Gammon, 2015, General Chemistry, Cengage, P.111
[11] Brown, LeMay, Burstein, Murphy, Woodward, Stottzfus, 2017, Chemistry the Central Science, Pearson, P.93
[12] Brown, LeMay, Burstein, Murphy, Woodward, Stottzfus, 2017, Chemistry the Central Science, Pearson, P.94–95
[13] https://pubchem.ncbi.nlm.nih.gov/compound/Glycerol
[14] Brown, LeMay, Burstein, Murphy, Woodward, Stottzfus, 2017, Chemistry the Central Science, Pearson, P. 102.

[15] Brown, LeMay, Burstein, Murphy, Woodward, Stottzfus, 2017, Chemistry the Central Science, Pearson, P.106

[16] Brown, LeMay, Burstein, Murphy, Woodward, Stottzfus, 2017, Chemistry the Central Science, Pearson, P.703

[17] Brown, LeMay, Burstein, Murphy, Woodward, Stottzfus, 2017, Chemistry the Central Science, Pearson, P.1048

[18] Chegeni A, Babaeipour V, Fathollahi M, Hosseini SG. Development of a new formulation of air revitalization tablets in closed atmospheres using the Taguchi statistical method. International Journal of Environmental Science and Technology. 2021;13:1–6.

[19] Chen KH, Miao YB, Shang CY, Huang TY, Yu YT, Yeh CN, Song HL, Chen CT, Mi FL, Lin KJ, Sung HW. A bubble bursting-mediated oral drug delivery system that enables concurrent delivery of lipophilic and hydrophilic chemotherapeutics for treating pancreatic tumors in rats. Biomaterials. 2020;255:120157.

6 The Effect of Lack of Oxygen and Excess Carbon Dioxide Buildup on Joint Diseases and the Natural Treatments for Some Joint Diseases

6.1 Overview

The leading cause of disability worldwide is arthritis. Arthritis can be either an auto-immune joint disease or osteoarthritis (OA). According to the Center for Disease Control (CDC), by 2040, 80 million US adults will have some form of arthritis. This chapter presents the effect of the obstructed ability to breathe by a physical block limit or by prolonged hypoventilation or by a pulmonary condition on hypoxia, hypercapnia, and blood chemical pH imbalance. Mechanism of alternating blood pH values, methods of identifying blood chemical pH imbalance, methods of measuring P_{CO_2}, P_{O_2}, and blood pH, types of joint diseases caused due to blood chemical pH imbalance, current medicine prescribed in their treatments, and the natural treatments of such joint diseases are also discussed.

6.2 Introduction and Background

The earth's atmosphere contains 21% oxygen. A closed space has safe oxygen levels if readings are between 20.8 and 21%, while a space with readings of less than 19.5% is oxygen-deficient according to OSHA guidelines. Home air conditioners: split AC, window AC, and portable AC can't ventilate your room: they don't have the capability to bring the outside air; they only circulate indoor air and cool it. Complex HVAC systems used inside some hotels, office buildings, big malls, and airports have the feature of ventilating the indoor area and maintaining freshness, humidity, and oxygen level. Some window ACs in the US can bring outside air and maintain oxygen levels. Most window ACs around the globe do not have this feature. Oxygen room level of 19.5–23.5% is considered safe, and oxygen levels between 14 and 16% and below 19.5% are considered hazardous.

If, while breathing out, not enough carbon dioxide is expelled from the lungs, the increased carbon dioxide in the blood reduces the blood's pH and makes the blood acidic, causing respiratory acidosis. When your ability to breathe is blocked by physical block limits, another condition or a disease, such as respiratory acidosis, is caused. Respiratory acidosis can be either acute, chronic, or acute and chronic. The sudden arrival of CO_2 in the lungs is called acute respiratory acidosis. The kidney's response to acute respiratory acidosis is so quick that it can happen within minutes. The causes may include cerebrovascular accidents like strokes, a group of diseases that interfere with gene's ability to make muscle and causes steady loss of muscles causing muscu-

https://doi.org/10.1515/9783111316864-006

lar dystrophy, voluntary muscles may become weak, or loss of control of them causing myasthenia gravis, heart attack, a very rare neurological disorder where the immune system attacks itself and can cause problems from trouble eating to full body paralysis causing Guillain-Barre syndrome and/or block airways.

Chronic respiratory acidosis is more serious, and it happens at a slower rate and at a lesser degree than acute respiratory acidosis. The lower oxygen rate for the tissues to be fully supplied in the blood is known as hypoxemia. The conditions may include pulmonary fibrosis or diseases that happens in the lung tissue/nerve or muscular diseases, sleep apnea, obesity, and thoracic skeletal defects that cause pecs/rib cage/or sternum to be shaped in a way that limits lung functioning or breathing, asthma, and a group of airflow and breathing diseases including diseases like bronchitis and emphysema called chronic obstructive pulmonary disease (COPD). Common treatments for respiratory acidosis are oxygen tubes, different medications or other treatments to stop smoking, naloxone (for opioid overdose), anti-inflammatory medications to ease any constrictive swelling, and breathing machines like a continuous positive airway pressure (CPAP) or BiPAP. To prevent getting respiratory acidolysis in general, lose weight if overweight, quit smoking, and don't drink alcohol while taking opioids and strong pain medications.

A 2-year study showed that there is an adverse physical effect on medical staff when wearing an N95 mask. Wearing an N95 mask resulted in hypercapnia (excessive CO_2 in the blood caused by inadequate respiration) and hypoxemia (lack of oxygen in the blood) which reduced the ability to make a correct decision and the working efficiency. Dizziness, shortness of breath, and headache were experienced by the medical staff wearing N95 masks.

If prolonged hypoventilation accompanies respiratory acidosis, the condition becomes more severe, and it can cause the patient to have additional symptoms: myoclonus, seizures, and altered mental status. Hypercapnia (excessive amounts of carbon dioxide in the blood) can be caused by respiratory acidosis leading to cerebral vasodilation. Severe respiratory acidosis may cause papilledema and increased intracranial pressure increasing the risk of death and herniation. Chronic respiratory acidosis may cause impaired coordination, polycythemia, memory loss, heart failure, and pulmonary hypertension.

Dyspnea (difficult respiration) is commonly caused by COPD, interstitial lung disease, heart failure, asthma, psychogenic problems that are usually linked to anxiety, and pneumonia. Chest X-rays and computed tomography (CT) images are used by doctors for its diagnosis. Spirometry tests can be used to measure the airflow and the patient's lung capacity and to pinpoint the extent and the type of an individual's breathing problems. Other tests can be used to directly measure the blood capacity to carry oxygen and the level of oxygen in the blood. Treatment of dyspnea is dependent on its cause. If it is caused from having bacterial pneumonia, antibiotics are prescribed. If it is caused from having asthma, bronchodilators and steroids are prescribed, and other medications are also effective (opiates, antianxiety drugs, and nonsteroidal anti-inflammatory drugs (NSAIDs)). Special breathing techniques, such as breathing muscle strengthening exercises and pursed-lip breathing, can be used if the cause of dyspnea is caused by COPD. Supplemental oxygen may be prescribed if tests indicate low levels of oxygen in the blood.

A less well-known side effect of using NSAIDs (drugs) is the degradation of joint cartilage. Nonsteroidal anti-inflammatory drugs include: ibuprofen, used in some drugs such as mortin, nuprin, and advil; piroxicam, used in drugs such as feldene; diclofenac, used in drugs such as voltaren; fenoprofen, used in drugs such as nalfon; indomethacin used in drugs such as indocin; naproxen used in drugs such as naprosyn; tolmetin used in drugs such as tolectin; and sulindac used in drugs such as clinoril. Other side effects of using nonsteroidal anti-inflammatory drugs are ulcer formation, dizziness, headaches, and gastrointestinal upset. Clinical studies have shown that nonsteroidal anti-inflammatory drug usage has caused an acceleration of OA and increased joint destruction.

Common joint diseases may include bursitis, OA, gout, rheumatoid arthritis (RA), spondyloarthritis, lupus, and juvenile idiopathic arthritis. Bursitis is caused by the inflammation of the small, fluid-filled sacs called bursae around the joints, tendons, muscles, and bones. Overuse or sudden injury of joints such as the elbow, hip, and shoulder can lead to flare-ups. Bursitis can sometimes result from bacterial infections. OA is the wear-and-tear form that increases with age. Adults in their 50s and older are more likely to develop this kind of chronic and progressive disease. Women are more vulnerable to OA than men. It is stiffness and pain with movement caused by break down of cartilages that cushion the joints. The flexibility decreases, and walking becomes more difficult, especially with knee and hip arthritis. The type of arthritis that affects the joint connecting the big toe to the rest of the foot is called gout. A waste product in the blood, uric acid, would exist in excess and forms crystals in the joints. Flare-ups caused by gout are extremely painful and it would commonly strike at night. Men are more vulnerable to gout disease, and women become more vulnerable to this disease after menopause.

The autoimmune condition that affects the lining of the joints is called RA. Immune system cells accumulate in large numbers in the joints. The interaction between joint cells and immune cells causes increased inflammation leading to damage and destruction of the bone and cartilage. Spondyloarthritis consists of certain other rheumatoid diseases, including axial spondylitis, enteropathic arthritis, and psoriatic arthritis. The inflammation in the spine that can eventually lead to spinal fusion or ankylosing spondylitis is called axial spondylitis. The complication of inflammatory bowel diseases, like ulcerative colitis, is called enteropathic arthritis. Psoriatic arthritis is associated with the skin condition psoriasis, and it tends to affect the joints of the hands and feet. The autoimmune condition affects various parts of the body, including internal organs, skin, brain, blood, bones, and joints are called lupus. Lupus can cause an inflammation that triggers arthritis, particularly in the knees, feet, hands, elbows, and shoulders. The most common chronic joint condition in kids is juvenile idiopathic arthritis. The child's immune system attacks the body's own healthy tissue, and it is an autoimmune condition. The cause is unknown, and it may alter children's normal growth. This inflammation may affect the internal organs, eyes, muscles, joints, and ligaments.

6.3 Review and Discussion

The amount of oxygen circulating in the blood is the blood oxygen level. The normal reading using an oximeter is between 95 and 100. Forms of carbon dioxide carried in the blood are carbaminohemoglobin (CO_2 bound to hemoglobin) and chemically modified bicarbonate (HCO_3^-). The solubility of CO_2 in the blood is 0.07 mL CO_2/100 mL blood/mm Hg which would be almost 5% of the total CO_2 content of the blood. The solubility of oxygen in the blood is 0.003 mL O_2/100 mL blood/mm Hg, which would be almost 2% of the total O_2 content of the blood. The solubility of carbon dioxide is 20 times more soluble in blood than oxygen. According to Henry's law, the solubility of a gas is directly proportional to the partial pressure of that gas above the liquid. Increasing the pressure and decreasing the temperature would lead to an increase in the solubility of gaseous over a liquid [16]. CO_2 gas has a lower ability to diffuse and exit the lungs compared to oxygen gas according to Graham's law [95]. As it is denser than oxygen gas, the density of oxygen gas is 1.43 g/L compared to the density of carbon dioxide gas 1.81 g/L.

The body normally maintains CO_2 in a range from 38 to 42 mm Hg by balancing its elimination and production. Ventilation is primarily initiated by the blood pH. The pH of the blood is mainly regulated by the amount of CO_2 in the blood. The body produces more CO_2 than it can eliminate in case of hypoventilation, causing retention of CO_2. The increased CO_2 is what leads to an increase in blood acidity due to an increase of hydrogen ion concentration, and a slight increase in bicarbonate concentration, and the equilibrium would shift toward forming more hydrogen ions according to the following reaction:

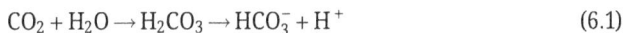

$$CO_2 + H_2O \rightarrow H_2CO_3 \rightarrow HCO_3^- + H^+ \tag{6.1}$$

A buffer system is created from the presence of the flowing molecules: HCO_3, CO_2, and $H_2CO_3^-$ in equilibrium.

In the presence of excess hydroxide ions (OH−), carbonic acid (H_2CO_3) would buffer a high pH, and in the presence of excess hydrogen ions (H+), carbonate anion (HCO_3^-) would buffer a low pH which is the main mechanism behind respiratory acidolysis blood pH drop. In respiratory acidolysis, the slight increase in bicarbonate acts as a buffer for the increase in H^+ ions, which helps minimize the drop in pH blood value. Increased hydrogen ions (H^+) would lead to a slight decrease in the buffered blood pH; blood pH would be below 7.35.

To evaluate patients with suspected respiratory acidosis, serum bicarbonate level and an arterial blood gas (ABG) are necessary. An elevated bicarbonate level of HCO_3^-

(>30 mmHg), an elevated PCO_2 (>45 mmHg), and decreased pH (<7.35) would show on an ABG test in case of respiratory acidolysis. Respiratory acidosis can be either chronic or acute based on the relative increase in HCO_3^- with respect to PCO_2. An HCO_3^- will have increased by one mEq/L for every 10 mmHg increase in PCO_2 over a few minutes in case of acute respiratory acidosis. An HCO_3^- will have increased by 4 mEq/L for every 10 mmHg increase in PCO_2 over a time course of days in case of chronic respiratory acidosis. A mixed respiratory-metabolic disorder may be present if it doesn't show either pattern of acute or chronic respiratory acidosis. A drug screen may also be warranted in a patient who shows unexplained respiratory acidosis.

The cause of respiratory acidosis must be treated once the diagnosis has been made. The rapid alkalization of the cerebrospinal fluid may lead to seizures; therefore, the hypercapnia should be corrected gradually. To help improve ventilation, pharmacologic therapy may be used. Beta-agonists, anticholinergic drugs, and methylxanthines (bronchodilators) may be used in treating patients with obstructive airway diseases. In case of patients who overdose on opioid use, naloxone can be prescribed.

Acidolysis can be either respiratory or metabolic in origin, depending on the measured pCO_2. If pCO_2 is greater than 40–45, it is known to be due to decreased ventilation, and it is called respiratory acidosis. If the pCO_2 is less than 40 since it is not the cause of the primary acid-base disturbance, it is called metabolic acidosis, and it is confirmed by a measured decrease in bicarbonate (normal range 21–28 mEq/L).

In alkalosis, the fluids of the body are alkaline; the blood pH is high. When the blood has too little acid making it basic, the condition is called alkalosis; blood pH would be higher than the normal pH value of 7.45. There might be no noticeable symptoms for mild and chronic alkalosis. If there is a rapid pH increase, symptoms may include confusion, nausea or vomiting, muscle twitching or spasms, numbness of the hands and feet, and/ or dizziness or lightheadedness.

Alkalosis can be either respiratory or metabolic, depending on pCO_2. If pCO_2 is greater than 45 mmHg, it is called metabolic alkalosis, and the measured bicarbonate is greater than 29 mM. If pCO_2 is less than 32 mmHg, it is called respiratory alkalosis, and the measured bicarbonate is less than 22 mM.

	pH	H^+	Pco_2	HCO_3^-
Normal	7.4	40 mEq/L	40 mm Hg	24 mEq/L
Respiratory acidosis	↓	↑	↑↑	↑
Respiratory alkalosis	↑	↓	↓↓	↓
Metabolic acidosis	↓	↑	↓	↓↓
Metabolic alkalosis	↑	↓	↑	↑↑

Fig. 6.1: Types of blood alkalosis and acidosis.

A number of diseases, including RA, are caused by alterations in tissue oxygen pressure. In a condition known as hypoxia, low partial pressure of oxygen is involved in angiogenesis, apoptosis, cartilage degradation, inflammation, oxidative damage, and energy metabolism. Synovial hypoxia can be linked to pathogenic processes through indirect and direct effects on angiogenesis, oxidative damage, inflammation, cartilage damage, and bone resorption. Studies show that hypoxia and other promoters lead to inflammation in RA. The metabolic environment in the synovium is modified by hypoxia, and an autoimmune response is initiated by the presentation of the upregulated antigenic enzymes in the context of cellular stress. Hypoxia induces an anaerobic glycolytic phenotype in the synovium. Several of the enzymes induced by this metabolic shift may become antigenic once anaerobic glycolysis is established in the synovium.

The abnormal biomechanics, attendant tissue-derived, and cell-derived factors cause OA. The progression of OA is related to reactive oxygen species (ROS) and oxidative stress. It is a multifactorial and polygenic joint disease. ROS target the complex oxidative stress signaling pathways as it regulates chondrocyte senescence and apoptosis, extracellular matrix synthesis and degradation, along with synovial inflammation and dysfunction of the subchondral bone and intracellular signaling processes. OA progresses from silent cartilage destruction to painful presentation. Free radicals mediate and amplify the sequence of joint degeneration in all tissues affected due to their chemical properties. Free radicals are the crucial factor involved in the inflammatory transformation of OA joints and all joint tissues disease development. Both OA and RA are caused by reduced oxygen levels from increased consumption by inflammatory cells such as synoviocytes and the oxygen-reduced delivery to synovial fluid.

Gout is a result of the presence of uric acid crystals building up in the joints. The production of uric acid in the joints is increased by the presence of excess carbon dioxide and the lack of presence of oxygen in the lungs. While sleeping, it produces less cortisone, an inflammation-suppressant; the reduction of cortisone level might be contributing to gout disease. A person may have as much as a 50% chance of getting gout disease if they suffer from sleep apnea. Dehydration and loss of water in the body can contribute to the increase in the concentration of uric acid in joint fluids, enhancing the formation of uric acid crystals in the joints and causing gout attacks. Gout is a type of arthritis that is caused by a uric acid concentration increase in the blood; it may cause debilitation due to uric acid deposits around tendons and joints. It is the most controllable metabolic disease. Gout is classified into primary and secondary. Primary gout causes are unknown. There are known genetic defects causing elevated uric acid. The increase of uric acid in primary gout can be due to the reduced ability to excrete uric acid found in a smaller group of patients (30%), increased formation of serum uric acid found in most patients, or both, which is found in a minority of patients.

RA can't be cured by drugs, but it can be treated as it can/will come back. Treatment of RA includes using medications to slow the progression of the disease. Drugs including sulfasalazine (azulfidine), methotrexate (trexall), and other biologic drugs such as etanercept (enbrel), and adalimumab (humira) may be prescribed. Biologic drugs reduce inflammation by targeting the immune system. Short-term treatment may include low-dose steroids. RA is a multifactorial disease as both environmental and genetic factors contribute to the disease. Medical therapy is limited in most RA cases; it fails to address the causes of the disease. As in OA, the use of NSAIDs drugs, including aspirin, is accompanied by the acceleration of factors that promote the disease process. Examples of drugs currently in use are hydroxychloroquine, penicillamine, methotrexate, gold therapy, azathioprine, and cyclophosphamide. A diet rich in vegetables, fiber, and whole foods and low in meat, sugar, saturated fat, and refined carbohydrates (Western diet) prevents the development of RA disease. Dietary therapy is to follow a vegetarian diet, eliminate food allergies, increase the intake of antioxidants, and alter the intake of fats and oils. Dietary therapy shows tremendous promise in the treatment of RA. Incomplete digestion may be a major factor in RA. Gamma-linolenic acid (GLA) acts as a precursor to an anti-inflammatory prostaglandins' series [1]. Studies show that some patients have responded to GLA treatment while others didn't. Fish oil supplementation containing omega-3 fatty acids shows a better and more positive response than GLA in the treatment of RA. Due to the neutralization of inflammation and support of collagen structure, dietary antioxidants such as flavonoids are used in the treatment of RA. Patients with RA have low levels of selenium. Selenium combined with vitamin E had a positive effect in the treatment of RA. Zinc levels are commonly low in RA patients; treatments with zinc supplements in the form of sulfates showed a slight therapeutic effect. Patients with RA are deficient in manganese-containing superoxide dismutase (SOD). The injectable form of the enzyme (antioxidant enzyme SOD (manganese SOD) available in Europe is effective in the treatment of RA. Oral administration of SOD showed no effect. Patients with RA are also deficient in vitamin C. Supplements with vitamin C give some anti-inflammatory action. Pantothenic acid in blood has been shown to be lower in RA patients. Patients who received 2 g of calcium pantothenate daily showed improvement. Arthritis patients showed a lower sulfur content in the fingernails. Using injectable sulfur alleviated pain and swelling. A high dose of niacinamide (900–4,000 mg) has shown good results in the treatment of both OA and RA. The administration of 500 mg of pantothenic acid has shown no effect on the treatment of RA.

A vegetarian diet following short-term fasting showed a reduction of RA disease activity.

Treatment of OA includes prescribing medications called bisphosphonates: risedronate (actonel) and alendronate (fosamax). In most OA cases, drug treatment has shown to be ineffective, and if the failure of nonsurgical treatment is consistent in at least three to six months, surgical replacement of large joints, knee replacement, or hip replacement, is needed.

The therapeutic goal in the natural treatment of OA is to enhance and repair the collagen matrix and the regeneration of the connective tissues. It is recommended to lose excess weight, causing increased stress on joints, use a healthy diet rich in complex carbohydrates and fiber, and minimize and eliminate the consumption of nightshade vegetables. Nightshade vegetables, including potatoes, tomatoes, peppers, tobacco, and eggplants, are known to contain alkaloids that promote the inflammatory degradation of the joints and inhibit normal collagen repair. The high intake of antioxidants is shown to inhibit the progression of the disease and reduce the risk of cartilage loss. As some people age, they lose their ability to manufacture sufficient levels of glucosamine. Glucosamine, in the form of glucosamine sulfate drug, is used to incorporate sulfur into the cartridges and stimulate the manufacture of glucosamineglycans.

Chondroitin sulfate is a drug that is composed of repeating units of derivatives of glucosamine sulfate attached to sugar molecules and is known to be a less effective drug than that of glucosamine sulfate due to its low absorption 0–13% compared to the solubility of glucosamine sulfate which is 90–98%. A high dose of niacinamide (900–4,000 mg) has shown good results in the treatment of both OA and RA. SOD injections showed a significant effect in the treatment of OA. Vitamin E has the ability to stimulate the formation of new cartilage components and inhibit the breakdown of cartilage, and the administration of 600 IU showed significant benefits. Vitamin C is like vitamin E and protects and enhances cartilage formation. The administration of a small amount (12.5 mg) of pantothenic acid is effective in relieving symptoms of OA. Joint degradation is accelerated by the deficiency of one of the vitamins: A, B6, zinc, copper, and boron, and supplementation at the appropriate level may promote cartilage synthesis and repair. Niacinamide in high dosage of 900–4,000 mg per day showed a promising result for the treatment of OA.

Treatments of gout joint disease include prescribed medications such as allopurinol and febuxostat. Treatment of sleep apnea that would include CPAP machine or another treatment device to increase oxygen intake while sleeping is used to increase oxygen level and lower uric acid production and reduce the risk of gout attacks. Decreasing the concentration of uric acid by drinking fluids would increase blood volume and lowers the risk of gout attacks. Other lifestyle changes that may lower the risk of a gout attack include getting regular exercise, eating a plant-based diet that is low in purines and whole foods, and losing excess weight. The standard medical treatment for the disease is the administration of colchicine, indomethacin, naproxen, fenoprofen, or phenylbutazone. The dietary treatment involves fluid intake, low fat, low purine, and low protein intake, consumption of complex carbohydrates, elimination of alcohol intake, and achieving the ideal body weight. The natural treatment of gout disease includes the consumption of nutritional supplements: eicosapentaenioc acid, vitamin E, folic acid, and amino acids such as alanine, aspartic acid, glutamic acid, glycine, and niacin, and vitamin C. Eicosapentaenioc acid and omega-3 oils are found useful in the treatment of gout joint disease. Vitamin E acts as an antioxidant and inhibits the formation of leukotrienes. Folic acid is known to inhibit the production of uric acid by inhibiting the enzyme xanthine oxidase, alanine, aspartic acid, glutamic acid, and glycine. These amino acids are shown to lower serum uric acid level by increasing uric acid excretion. Niacin and vitamin C, high doses of vitamin C, and niacin are used in the treatment of gout; niacin competes with uric acid in excretion, and vitamin C increases the formation of uric acid in small groups of people.

6.4 Conclusion

If the ability to breathe is obstructed by a pulmonary condition or by physical block limits or prolonged hypoventilation, respiratory acidosis is caused. Three main types of joint diseases are directly related to hypoxia (low levels of oxygen in your body tissues), hypercapnia (excessive amounts of carbon dioxide in the blood), and blood pH imbalance: RA, OA, and gout joint diseases.

Both OA and RA are caused by reduced oxygen levels from increased consumption by inflammatory cells such as synoviocytes and the oxygen-reduced delivery to synovial fluid. RA is directly caused by alterations in tissue oxygen pressure. OA is directly related to aging, as the ability to synthesize and restore cartilage structure decreases, and the progression of the disease was found to be directly related to the presence of ROS and oxidative stress. Gout joint disease is directly caused by the presence of uric acid crystals building up in the joints, and the crystal buildup is increased by the presence of excess carbon dioxide.

The natural treatment of both RA and OA includes minimizing the consumption of nightshade vegetables and eliminating the use of NSAIDs (drugs) as they cause the degradation of joint cartilages. A high dose of niacinamide (900–4,000 mg) has shown good results in the treatment of both OA and RA. Vitamin C is, like vitamin E supplements, has shown good results in the treatment of OA and RA and gout joint diseases. It is recommended to increase oxygen intake while sleeping and increase blood volume by drinking fluids to lower the risk of a gout attack. It is recommended to lose excess weight and maintain a healthy body weight in the treatment of both OA and gout joint diseases.

The dietary treatment of RA includes following a vegetarian diet, eliminating food allergies, increasing the intake of antioxidants, and altering the intake of fats and oils. The dietary treatment of OA includes the use of a healthy diet rich in complex carbohydrates and fiber to minimize and eliminate the consumption of nightshade vegetables and to intake antioxidants. The dietary treatment of gout joint disease includes fluid intake, low fat, low purine, and low protein intake, consumption of complex carbohydrates, and elimination of alcohol intake.

The natural treatment of gout joint disease includes the administration of nutritional supplements: eicosapentaenioc acid, vitamin E, folic acid, amino acids such as alanine, aspartic acid, glutamic acid, glycine, and niacin, and vitamin C. The natural treatment of RA disease includes the administration of some nutritional supplements: omega-3 fatty acid, selenium combined with vitamin E, vitamin C, niacinamide, and pantothenic acid. The natural treatment of OA disease includes the administration of nutritional supplements: glucosamine sulfate, niacinamide, vitamin E, vitamin C, small amounts of pantothenic acid, and niacinamide.

Chapter Questions

1) What's the aim of the chapter?
2) What's the amount of oxygen in air? What amounts are considered deficient?
3) What kind of diseases are caused by the effect of the obstructed ability to breathe?
4) What's the main cause of gout joint disease?
5) What's recommended in the natural treatment of Rheumatoid Arthritis, Osteoarthritis, and gout joint diseases?

References

[1]   Global Climate Change, "10 interesting things about air", September 2016, www.climate.nasa.gov. Retrieved May 2022.
[2]   GDS Team, "Understanding Safe Oxygen Levels as Outlined by OSHA in Confined Spaces", May 2017, www.gdscorp.com. Retrieved May 2022.
[3]   Home Particle, "How is Oxygen level maintained in Air-Conditioned Room", November 2021, www.homeparticle.com. Retrieved May 2022.
[4]   Brennan D, "What's Respiratory Acidosis?", www.webmd.com, June 2021. Retrieved May 2022
[5]   US National Library of Medicine, "The Physiological Impact of N95 Masks on Medical Staff", www.clinicaltrials.gov, 2005. Retrieved May 31, 2022.
[6]   Contreras M, Masterson C, Laffey JG. Permissive hypercapnia: what to remember. Curr Opin Anaesthesiol. 2015 Feb;28(1):26–37. [PubMed].
[7]   Medical NEWS Today, "What's dyspnea?", www.medicalnewstoday.com, Medical News Today, 2018. Retrieved May 2022.
[8]   M.J. Sheild, "Anti-Inflammatory Drugs and their Effects on Catlilage Synthesis and Renal Function", Eur J Rheumatol Inflam 13 (1993):7–16.
[9]   P.M. Brooks, S.R. Potter, and W.W. Buchanan, "NSAID and Osteoarthritis – Help or Hinderance", J Rheumatol 9 (1982): 3–5.
[10] N.M Newman, and R.S.M. Ling, "Acetabular Bone Destruction Related to Non-steroidal Anti-Inflammatory Drugs", Lancet 2 (1985): 11–13.
[11] L. Solomon, "Drug-Induced Anthropology and Necrosis of the Femoral Head", J Bone Joint Surg 55B (1973): 246–51.
[12] H. Ronningen, and N. Langeland, "Indomethacin Treatment in Osteoarthritis of Hip Joints", Acta Orthop Scand 50(1979): 169–74.
[13] Lisa Esposito, "A Patient's Guide to Bone and Joint Diseases", www.health.usnews.com, U.S. News, June 2019. Retrieved May 2022.
      A. Pristas, "Blood Oxygen Level: What's all the Hype about?" https://www.hackensackmeridian health.org/, September 2002. Retrieved May 2022.
[14] Stanfield, C. L., Principles of Human Physiology 5th edition. Pearson, October 2012.

[15]   Henry, W. (1803). "Experiments on the quantity of gases absorbed by water, at different temperatures, and under different pressures". Phil. Trans. R. Soc. Lond. 93: 29–43. doi:10.1098/rstl.1803.0004.

[16]   Brinkman JE, Toro F, Sharma S. StatPearls [Internet]. StatPearls Publishing; Treasure Island (FL): Aug 24, 2021. Physiology, Respiratory Drive. [PubMed].

[17]   Kisaka T, Cox TA, Dumitrescu D, Wasserman K. CO_2 pulse and acid-base status during increasing work rate exercise in health and disease. Respir Physiol Neurobiol. 2015 Nov;218:46–56. [PubMed].

[18]   Katalinić L, Blaslov K, Pasini E, Kes P, Bašić-Jukić N. [Acid-base status in patients treated with peritoneal dialysis]. Acta Med Croatica. 2014 Apr;68(2):85–90. [PubMed].

[19]   Marhong J, Fan E. Carbon dioxide in the critically ill: too much or too little of a good thing? Respir Care. 2014 Oct;59(10):1597–605. [PubMed].

[20]   Cove ME, Federspiel WJ. Veno-venous extracorporeal CO_2 removal for the treatment of severe respiratory acidosis. Crit Care. 2015 Apr 17;19:176. [PMC free article] [PubMed].

[21]   Sharma S, Hashmi MF, Burns B. StatPearls [Internet]. StatPearls Publishing; Treasure Island (FL): Aug 30, 2021. Alveolar Gas Equation. [PubMed].

[22]   MacKenzie Burger and Derek J. Schalle, "Metabolic Acidolysis", National Library of Medicine, www.ncbi.nlm.nih.gov, 2021. Retrieved May 2022.

[23]   Expert Board, "Acidolysis and Alkalosis", www.testing.com, January 2022. Retrieved May 2022.

[24]   J. G. Betts, K. A. Young, J.A Wise, E. Johnson, B. Poe, D. H. Kruse, O. Korol, J.E. Johnson, M. Wombe, and P. Desaix, 2019, Anatomy and Sociology, OpenStax, http://cnx.org/contents/14fb4ad7-39a1-4eee-ab6e-3ef2482e3e22@15.5.

[25]   J. G. Betts, K. A. Young, J.A Wise, E. Johnson, B. Poe, D. H. Kruse, O. Korol, J.E. Johnson, M. Wombe, and P. Desaix, 2019, Anatomy and Sociology, OpenStax, http://cnx.org/contents/14fb4ad7-39a1-4eee-ab6e-3ef2482e3e22@15.5.

[26]   Quiñonez-Flores, C. M., González-Chávez, S. A., Pacheco-Tena, C., Hypoxia and its implication to Rheumatoid Arthritis, J Biomed Sci. 2016; 23(1): 62, doi: 10.1186/s12929-016-0281-0

[27]   Zahan O.M., Serban O., Gherman C., Fodor D, The evaluation oxidative stress in osteoarthritis, Med Pharm Rep. 2020 Jan; 93(1): 12–22.

[28]   Neogi T, Chen C, Niu J, et al. Relation of temperature and humidity to the risk of recurrent gout attacks. Am J Epidemiol. 2014;180(4):372–377. DOI: 10.1093/aje/kwu147

[29]   Roddy E. Revisiting the pathogenesis of podagra: why does gout target the foot? J Foot Ankle Res. 2011;4(1):13. Published 2011 May 13. PMID: 21569453 doi:10.1186/1757-1146-4-13

[30]   Martillo MA, Nazzal L, Crittenden DB. The crystallization of monosodium urate. Curr Rheumatol Rep. 2014;16(2):400. PMID: 24357445 doi:10.1007/s11926-013-0400-9

[31]   Choi HK, Niu J, Neogi T, Chen CA, Chaisson C, Hunter D, Zhang Y. Nocturnal risk of gout attacks. Arthritis Rheumatol. 2015 Feb;67(2):555–62. PMCID: PMC4360969 DOI: 10.1002/art.38917

[32]   Abhishek A, Valdes AM, Jenkins W, Zhang W, Doherty M. Triggers of acute attacks of gout, does age of gout onset matter? A primary care-based cross-sectional study. PLoS One. 2017;12(10):e0186096. Published 2017 Oct 12. PMID: 29023487 doi:10.1371/journal.pone.0186096

[33]   Bouloukaki I, et al. (2014). Intensive versus standard follow-up to improve continuous positive airway pressure compliance. European Respiratory Journal, 44(5): 1262–1274. PMID: 24993911 DOI: 10.1183/09031936.00021314.

[34]   R. Jenkins, P. Rooney, and D. Jones et al., "Increased Intestinal Permeability for Patients with Rheumatoid Arthritis: A Side Effect of Oral Non-Steroidal Anti-Inflammatory Drug Therapy ", Br J Rheumatol 26 (1987): 103–7.

[35]   Murray M, Pizzorno J., Encyclopedia of Natural Medicine 2nd edition, 1998. ISBN:978-0-7615-1157-1

[36]   L. G. Darlington and N.W. Ramsey, "Clinical Review: Review of Dietary Therapy in Rheumatoid Arthritis ", Br J Rheumatol 30 (1993): 507–14.

[37]   H. M. Buchanan, S.J. Preston, and P.M. Brooks et al., "Is Diet Important in Rheumatoid Arthritis ", Br J Rheumatol 30(1991): 125–34.

[38]   F. McCrae, K. Veerapen, and P. Dieppe, "Diet and Arthritis ", Practitioner 230 (1986): 359–61.

[39]   T. J. De Witte et al., "Hypochlorhydria and Hypergastrinemia in Rheumatoid Arthritis ", Ann Rheu Dis 38 (1979): 14–17.

[40]   K. Henriksson et al., "Gastrin, Gastric Acid Secretion, and Gastric Microflora in Patients with Rheumatoid Arthritis ", Ann Rheu Dis 45 (1986): 475–83.

[41]   M. Brzeski, R. Madhok, H.A. Capell, "Evening Primrose Oil in Patients with Rheumatoid Arthritis and Side Effects on non-Steroidal Ant-Inflammatory Drugs ", Br J Rheumatol 30 (1991): 371–2.

[42]   J. F. Blech et al., "Effects of Altering Dietary Essential Fatty Acids on Requirements for non- steroidal Anti-inflammatory Drugs in Patients with Rheumatoid Arthritis: A Double-Blinded Placebo-Controlled Study ", Ann Rheu Dis 47 (1988): 96–104.

[43]   L. J. Levanthal et al.,"Treatment of Rheumatoid Arthritis with Gamma-Linolenic Acid ", Annals Int Med 119 (1993): 867–73.

[44]   J. M. Kremer et al., "Fish Oil Supplementation in Active Rheumatoid Arthritis: A Double-Blinded, Controlled Cross-Over Study ", Ann Intern Med 106 (1987): 497–502.

[45]   R. Sperling et al., "Effects of Dietary Supplementation with Marine Fish Oil on Leukocyte Lipid Mediator Generation and Function in Rheumatoid Arthritis ", Arthritis Rheum 30 (1987): 988–97.

[46]   L. G. Cleland et al.,"Clinical and Biochemical Effects of Dietary Fish Oil Supplementation in Rheumatoid Arthritis ", J Rheumatol 15 (1988): 1471–5.

[47]   M. Magaro et al., "Influence of Diet with Different Lipid Composition on Neutrophil Chemiluminescence and Disease Activity in Patients with Rheumatoid Arthritis ", Ann Rheu Dis 47 (1988): 793–6.

[48]   H. Van Der Temple et al., "Effects of Fish Oil Supplementation in Rheumatoid Arthritis ", Ann Rheu Dis 49 (1990): 76–80.

[49]   J. M. Kremer et al., "Dietary Fish Oil and Olive Oil in Patients with Rheumatoid Arthritis", Arth Rheum 33 (1990): 810–20.

[50]   C.S Lau, et al., "Maxepa On Non-Steroidal Anti-inflammatory Drug Usage in Patients with Mild Rheumatoid Arthritis ", Br J Rheumatol 30 (1991): 137.

[51]   G. L. Nielsen et al., "The Effects of Dietary Supplementation with N-3 Polyunsaturated Fatty Acids in Patients with Rheumatoid Arthritis, A Randomized double-blinded trial", Eur J Clin Invest 22 (1992): 678–91.

[52]   V. Codey, E. Middleton, and J. B. Harborne, Plant Flavonoids in Biology and Medicine- Biochemical, Pharmacological, and Structure-Activity Relationships (New York: Alan R. Liss, 1986); V. Codey, E. Middleton, J. B. Harborne, and A. Beretz, Plant Flavonoids in Biology and Medicine II-Biochemical, Pharmacological, and Structure-Activity Relationships (New York: Alan R. Liss, 1988).

[53]   U. Tarp at el., "Low Selenium Level in Severe Rheumatoid Arthritis", Scandinavian Journal of Rheumatology 14(1985): 97–101.

[54]   E. Munthe, and J Aseth, "Treatment of Rheumatoid Arthritis with Selenium and Vitamin E", Scandinavian Journal of Rheumatology 53 (suppl), 1984:103.

[55]   S.P. Pandley, S.K Bhattacharya, and S. Sundar, "Zinc in Rheumatoid Arthritis ", Indian Journal of Medical Research 81(1985): 618–20.

[56]   P.A. Simkin, "Treatment of Rheumatoid Arthritis with oral Zinc Sulfate ", Agents and Actions (Suppl.), 8(1981):587–95.

[57]   P.C. Mattingly, and A.G. Mowat, "Zinc Sulfate in Rheumatoid Arthritis", Annals of the Rheumatic Diseases 41(1982): 456–7.

[58]   K. B. Menander-Huber, "Orgotein in the Treatment of Rheumatoid Arthritis ", Europ J Rheum inflammation 4(1981): 201–11.

[59] S. Zidenberg-Cherr et al., "Dietary Superoxide Dismutase Does not Affect Tissue Level ", Am J Clin Nutr 37(1983): 5–7.
A. Mullen and C.W.M. Wilson,"The Metabolism of Ascorbic Acid in Rheumatoid Arthritis", Proc. Nutr. Sci 35(1976): 8A–9A.

[60] N. Subramanian, "Histamine Degradation Potential of Ascorbic Acid ", Agents and Actions 8 (1978): 484–7.

[61] M. Levine, "New Concepts in Chemistry and Biochemistry of Ascorbic Acid ", New Engl J Med 314 (1986): 892–902.

[62] E.C. Barton-Wright, and W.A Elliott, "The Pantothenic Acid Metabolism of Rheumatoid Arthritis ", Lancet 2 (1963): 862–3.

[63] General Practitioner Research Group, "Pantothenic Acid in Rheumatoid Arthritis ": Practitioner 224 (1980): 208–11.

[64] B. D. Senturia,"Results of Treatment of Chronic Arthritis and Rheumatoid Conditions with Colloidal Sulfur", J Bone Joint Surg 16(1935): 185–8.

[65] K. Wheeldon, "The Use of Colloidal Sulfur in the Treatment of Arthritis ", J Bone Joint Surg 17 (1935): 693–726.

[66] K. Ransberger, "Enzyme Treatment of Immune Complex Diseases", Arthritis Rheuma 8 (1986): 16–19.

[67] J. A. Shapiro et al., "Diet and Rheumatoid Arthritis in Women: A Possible Protective Effects of Fish Consumption ", Epidemiology 7 (1996): 256–63.

[68] G. W. Comstock et al., "Serum Concentrations of Alpha Tocopherol, Beta Carotene, and Retinol Preceding the Diagnosis of Rheumatoid Arthritis and Systematic Lupus Erythematosus ", Ann Rheum Dis 56(1997):323–5.

[69] N.F. Childers and G.M. Russo, The Nightshades and Health (Summerville NJ: Horticulture Publications, 1973).

[70] T.E. McAlindon et al., "Do Antioxidant Micronutrients Protect against the Development and Progression of Knee Osteoarthritis?", Arthritis Rheumatism 39 (1996):648–56.

[71] K. Karzel, and R. Domenjoz, "Effect of Hexosamine Derivatives and Uronic Acid Derivatives on Glycoaminoglycans Metabolism of Fibroblast Cultures", Pharmacology 5 (1971): 337–45.

[72] R.R. Vidal y Plana et al., "Articular Cartilage Pharmacology: I. In vitro Studies on Glucosamine and Non-Steroidal Anti-inflammatory Drugs", Pharmacol Res Comm 10 (1978): 557–69.

[73] I Setnikar et al., "Pharmacokinetic of Glucosamine in Man", Arzneim Forsch 43 (10) (1993): 1109–13.
A. Biaci et al., Analysis of Glycoaminoglycans in Human Sera after Oral Administration of Chondroitin Sulfate", Rheumatol Int 12 (1992): 81–8.

[74] A Conte et al., "Biochemical and Pharmacokinetic Aspect of Oral Treatment with Chondroitin Sulfate", Arzneim Forsch 45 (1995): 918–25.

[75] W. Kaufman, The Common Form of Joint Dysfunction: Its Incidence and Treatment (Brattleboro, VT: EL Hildreth Co., 1949.
A. Hoffer, "Treatment of Arthritis by Nicotinic Acid and Nicotinamide", Can Med Assoc J 81(1959): 235–9.

[76] K. Lund- Olesen, and K.B. Menander,"Orgotein: A new Anti-inflammatory Metalloprotein Drug: Preliminary Evaluation of Clinical Efficacy and Safety in Degenerative Joint Diseases", Curr Ther Res 16 (1974): 706–17.

[77] E.C. Huskisson and J. Scott, "Orgotein in Osteoarthritis of the Knee Joint", Eur J Rheumatol Inflam 4 (1981): 212.
I. Machtey, and L. Quaknine, "Toco-phenol in Osteoarthritis: A Controlled Pilot Study", J Am Ger Soc 26(1978): 328–30.

[78] E.R. Schwartz, "The Modulation of Osteoarthritic Development by Vitamin C and E", Int J Vit Nutr Res Suppl 26 (1984): 141–46.

[79]   C.J. Bates, "Proline and Hydroxyproline and Vitamin C Status in Elderly Human Subjects", Clin Sci
        Mol Med 52 (1977): 535–43.
[80]   A.P. Prins, J.M. Lipman, C.A. McDevitt, and L. Sokoloff, "Effect of Purified Growth Factors on Rabbit
        Articular Chondrocytes in Monolayer Culture', Arthr Rheum 25 (1982): 1228–32.
[81]   J.C. Anand, "Osteoarthritis and Pantothenic Acid", J Coll Gen Pract 5, (1963):136–37.
[82]   J.C. Anand, "Osteoarthritis and Pantothenic Acid", Lancet 2, (1963):1168.
[83]   W. Kaufman, The Common Form of Joint Dysfunction: Its Incidence and Treatment (Brattleboro, VT:
        E.L. Hildreth Company, 1949.
        A. Hoffer, "Treatment of Arthritis by Nicotinic Acid and Nicotinamide", Canadian Medical Association
        Journal 81 (1959): 235–9.
[84]   Efthimiou, Petros. (2020). Absolute Rheumatology Review. 10.1007/978-3-030-23022-7.
[85]   Zhang Y, Peloquin CE, Dubreuil M, et al. Sleep Apnea and the Risk of Incident Gout: A Population-
        Based, Body Mass Index-Matched Cohort Study. Arthritis Rheumatol. 2015;67(12):3298–3302. PMID:
        26477891 doi:10.1002/art.39330
[86]   Giles TL, et al. (2006). Continuous positive airways pressure for obstructive sleep apnoea in adults.
        Cochrane Database of Systematic Reviews (3). PMID: 16437429.
[87]   DOI: 10.1002/14651858.CD001106.pub2
[88]   Neogi T, Chen C, Niu J, et al. Relation of temperature and humidity to the risk of recurrent gout
        attacks. Am J Epidemiol. 2014;180(4):372–377. DOI: 10.1093/aje/kwu147
[89]   Ball D., Key J., Introductory Chemistry 1st Canadian Edition, 2014, ISBN:978-1-77420-002-5.
[90]   CONCOA Gas Controls, 2012, "Oxygen (O2)", www.concoa.com. Retrieved May 2022.
[91]   Engineering ToolBox, (2018). *Carbon dioxide – Density and Specific Weight vs. Temperature and Pressure.*
        [online] Available at: https://www.engineeringtoolbox.com/carbon-dioxide-density-specific-weight-
        temperature-pressure-d_2018.html. Retrieved May 2022.
[92]   L. Skoldstam, L. Larsson, and F.D Lindstorm, "Effects of Fasting and Lactovegetarian Diet on
        Rheumatoid Arthritis", Scand J Rheumatol 8 (1979):137–44.
[93]   I. Hafstrom, et al., "Effects of Fasting on Disease Activity, Neutrophil Function, Fatty Acid
        Composition, and Leukotriene Rheumatoid Arthritis", Arthr Rheum 31 (1988): 585–92.
[94]   R. Petersdorf et al., eds., Harrison's Principles of Internal Medicine (New York: McGraw-Hill, 1983)
[95]   M.V. Krause and L.K. Mahan, Food, Nutrition, and Diet Therapy, 7th edition, (Philadelphia: W.B
        Saunders, 1984), 677–9.
[96]   Neutrition Foundation, Present Knowledge in Nutrition, 5th edition (Washington D.C.: Nutrition
        Foundation, 1884), 740–56.
[97]   F.X. Pi-Sunyer, "The Fattening of America", JAMA 272 (1994): 238.
[98]   R.V. Panganamala and D.G. Cornwell, "the Effect of Vitamin E on Arachidonic Acid Metabolism" Ann
        NY Acad Sci 393 (1982): 376–91.
[99]   A.S. Lewis, L. Murphy, C. MCalla, et al., "Inhibition of Mammalian Xanthine Oxidase by Folate
        Compounds and Amethopterin", J Biol Chem 259: 15–5, 1984.
[100]   A. Bindoli, M. Valentine, and L. Cavallini, "Inhibitory Action of Quercetin on Xanthine Oxidase and
        Xanthine Dehydrogenase Activity", Pharm Res Comm 17(1985): 831–9.
[101]   S.L. Gershon, and I.H. Fox, "Pharmacological Effects of Nicotinic Acid on Human Purine
        Metabolism", J Lab Clin Med 84 (1974):179–56.
[102]   Levick J.R. Hypoxia and acidosis in chronic inflammatory arthritis; relation to vascular supply and
        dynamic effusion pressure. *J Rheum.* 1990;17(5):579–582.
[103]   Biniecka M., Kennedy A., Fearon U., Ng C.T., Veale D.J., O'Sullivan J.N. Oxidative damage in synovial
        tissue is associated with in vivo hypoxic status in the arthritic joint. *Ann Rheum Dis.* 2010;69
        (6):1172–1178.

7 Enthalpy and Bond Dissociation Energy Values for Multifluorinated Ethanol and Its Radicals Using Gaussian M-062x/6-31+g (d,p) Method at Standard Conditions

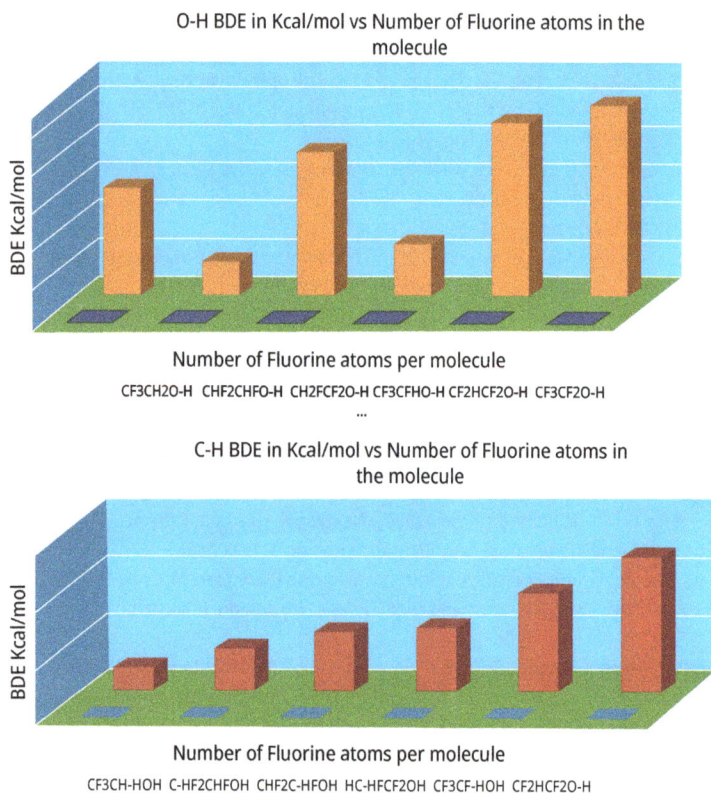

Fig. 7.1: C–H and O–H bond dissociation energy values and its relation to number of fluorine atoms per molecule.

7.1 Overview

Fluorinated alcohols are used as solvents for proteins, organic compounds, and peptides. They are also known to be used in the organic synthesis industry due to their strong hydrogen bonding character. Halogenated hydrocarbons are mostly synthetically produced, and they don't exist naturally in the environment. Bond dissociation energy (BDEs) values can be used to explain the reactivity and stability of chemical

https://doi.org/10.1515/9783111316864-007

compounds. This chapter presents the methods used to calculate some kinetic and thermodynamic parameters: standard enthalpy of formation and BDE values for 18 different fluorinated ethanol, tri-fluorinated ethanol, tetra-fluorinated ethanol, and penta-fluorinated ethanol using the popular ab initio and density functional theory Gaussian M-062x/6-31+g (d,p) method.

Optimized structures and thermochemical properties of tri-, tetra-, and penta-fluorinated ethanols and its radicals determined by the Gaussian M-062x/6-31 + g (d,p) calculation are presented in this chapter. The tabulated literature values for calculated enthalpies of formation and BDEs for 18 fluorinated ethanol and some radicals via several series of isodesmic reactions are also included.

7.2 Introduction

Fluorinated alcohols are known to be used in the organic synthesis industry. They have strong hydrogen bonding donor character, and they are strong nucleophiles that allow organic reactions to occur without the use of a catalyst. Fluorinated alcohols have been used as solvents in epoxidation reactions, annulation reactions, nucleophilic substitution reactions, electrophilic reactions, and the functionalization of multiple bonds.

Fluorinated alcohols are excellent solvents of proteins, peptides, and other organic compounds due to their physicochemical properties. Fluoro-alcohols, like other alcohols, can alter lipid bilayer properties and stability, and protein function.

Most halogenated hydrocarbons don't exist naturally in the environment and are synthetically produced. The major source of halogenated hydrocarbons in the atmosphere is agriculture. Crop spraying introduces halogenated hydrocarbons to the environment entering through adsorption and deposition onto airborne particles or directly entering the aquatic system. Some halogenated hydrocarbons, PCBs, furans, and dioxins are by-products of industrial waste that would unintentionally enter the atmosphere.

The onset temperatures for energetic materials in the calorimetric measurements have been roughly predicted using molecular orbital calculations of BDEs. The stability and reactivity of chemical compounds can be explained using BDEs values. Standard enthalpies of formation estimated using semiempirical MO calculations, the MOPAC-PM7 package, has been used previously to derive BDE values for chemical compounds.

BDEs values for some inorganic compounds, lanthanide selenides, and sulfides were measured in 2021 using resonant two-photon ionization spectroscopy. The predissociation thresholds were found to be the BDE values for these molecules. The 0 K gaseous heat of formation, ΔH_f, for each molecule, was also reported using this method.

The amount of energy used to break a mole of the covalently bonded gas molecule to a pair of radicals is the BDE. The units used for the BDE are commonly kilojoule per mole. Covalent bonds can be broken heterolytically or homolytically. The heterolytic breaking of a covalent bond would result in the pair of electrons going to only one atom, either A or B:

$$C - D \rightarrow C^+ + D^{:-} \text{ or } C - D \rightarrow C^{:-} + D^+.$$

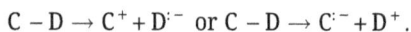

The homolytic breaking of a covalent bond, on the other hand, would result in one electron staying with each atom, $A - B \rightarrow A \bullet + B \bullet$. BDE can be calculated for molecules as the difference in enthalpy of the formation of products and reactants. BDE is a state function, as it doesn't depend on the mechanism or pathway of how bonds form or break. The energy of chemical reactions can be assessed using values for the BDE. There are some systematic trends for the bond dissociation values; BDE varies with hybridization. For example, sp^3-hybridized carbons in hydrocarbons have smaller bond dissociation values compared to sp^2-hybridized carbons. The longer and weaker sp^3-hybridized bond is easier to break compared to the shorter and stronger sp^2- and sp-hybridized bonds (double and triple bonds). Among sp^3-hybridized bonds, bond dissociation values depend on their position, whether it's on a primary, secondary, or tertiary carbon. Methane has the strongest C–H bond with the highest bond dissociation values, following C–H bonds on primary carbons, following C–H bonds on secondary carbons, and following C–H bonds on tertiary carbons.

Energetics of chemical processes can be assessed using BDEs. Hess's law has been used in the past and is currently being used to estimate reaction enthalpies by combining BDEs of bonds formed and BDEs of bonds broken.

The energy change when forming a mole of compound from its component elements is called the enthalpy of formation, ΔH_f. If heat is released when the elements combine to form the compound, enthalpy of formation would have a negative sign. If heat is absorbed when the elements combine to form the compound, the enthalpy of formation would have a positive sign. Values of the enthalpy of formation are dependent upon temperature, pressure, and physical states of reactants and products in the chemical reaction. Standard enthalpy of formation (ΔH_f^o) is the enthalpy of formation at standard conditions: 1 atm pressure, 25 °C, and 1 M aqueous solution concentration. Any element in its most stable form has a standard enthalpy of formation and has a value of zero. Tabulated enthalpy of formation values can be used to calculate the standard enthalpy of any reaction whose standard enthalpy of formation values are well known:

$$\Delta H_{rxn}^o = \sum m\Delta H_f^o(\text{products}) - \sum n\Delta H_f^o(\text{reactants})$$

The heat involved in chemical or physical change at constant temperature and pressure is the enthalpy of reaction (H), and its thermodynamic quantity $q = \Delta H$.

Molecules, ions, or atoms containing at least an unpaired electron in the valance shell are called free radicals. Free radicals are chemically reactive, unstable, and mostly short-lived. Heat, electrolysis, electrical discharge, and ionizing radiation can generate free radicals. Free radicals are intermediates in many chemical reactions. Free radicals are important in atmospheric chemistry, combustion, plasma chemistry, polymerization, and biochemistry, and they are important in many chemical reactions.

In 2016 Wang studied the thermodynamic properties of fluorinated methanol using CBS-QB3, M06-2X, WB97X, W1U, M06, B3LYP, CBS-APNO, and G4 calculations. A small standard deviation suggests good error cancellation of work reactions and accuracy. Small values for standard deviations were obtained in calculations using the M06-2x/6-31+g (d,p) Gaussian method; it is an accurate method to calculate the enthalpy of fluorinated alcohols; it shows the second smallest standard deviation after the CBS-QB3 method of calculation. The enthalpy of fluorinated methanol was studied in the past.

Halogenated compounds have low reactivity, are highly stable, and are used in industry. They are of concern to the environment due to their persistence in the environment and their widespread use. Their thermochemical properties must be studied in order to understand the reduction and oxidation reactions involving these molecules.

7.3 Experimental and Data

7.3.1 Computational Method

The Global-hybrid meta-generalized gradient approximation (GGA) density functional approximation, GGA, DFT Gaussian M-062x/6-31+g (d,p) method of calculation has been used to initially analyze frequencies, optimized structures, and thermo energies of the molecules studied. In the GGA, the density functional depends on the down and up spin densities and the reduced gradient. In the meta-GGA, the function also depends on the up and down spin kinetic energy densities. A hybrid GGA is a combination of GGA with the Hartree-Fock exchange. The hybrid meta-GGA is a combination of meta-GGA with the Hartree-Fock exchange.

7.3.2 Isodesmic and Isogyric Reaction

A series of isodesmic reactions and composite calculations were employed to calculate the enthalpy of the formation of tri-, tetra-, and penta-fluorinated ethanol. Calculations were performed using Gaussian 16 program. All reported calculations of enthalpy of formation are for standard conditions of 1 atm pressure and 298 K. The Gaussian M-062x/6-31+g (d,p) level of calculation has been used for this study as this method of calculation was successfully employed in the past when applied to fluoro hydrocarbon 6 with small reported standard deviation values.

The standard deviation is calculated using the following formula:

$$\sigma = \sqrt{\frac{1}{N} \sum_{i=1}^{N} (x_i - \mu)^2}$$

where X_i is the mean; the average of the numbers, μ is the actual numbers to be calculated the standard deviation of, and

$$\frac{1}{N}\sum_{i=1}^{N}(x_i-\mu)^2 \text{ is the variance.}$$

The Gaussian M-062x/6-31+g (d,p) method of calculation used to calculate the enthalpy of formation of tri-, tetra-, and penta-fluorinated ethanols and its radicals are presented. Work reactions and reference species used to calculate the enthalpy of formation of fluorinated ethanol using this method are also presented. The number of each type of bond must be conserved in each isodesmic reaction to cancel any systematic error in the molecular orbital calculations using this method. By careful choice of the isodesmic reactions, all enthalpies of formation calculations are allowed to obtain accuracy close to experimental values. The ΔH_{f298K° values of all reference species but the fluoroethanols are known; the $\Delta H_{f298°K}$ of the target species fluoro-ethanols is obtained from this data, and the calculated ΔH_{rxn}, 298°. $\Delta H_{f298°K}$ calculated using two different reference molecules are within $\pm(0-0.60 \text{ kcal mol}^{-1})$.

7.3.3 Reference Species

Table 7.1 lists the standard enthalpy of formation for the reference species used in isodesmic reactions with their uncertainties. Table 7.2 provides all calculated standard enthalpy of formation values, $\Delta_f H°_{(298)}$ for tri-, tetra-, and penta-fluorinated ethanol and its radicals.

Tab. 7.1: Reference species in the isodesmic reactions standard enthalpy of formation values (kcal mol^{-1}).

Species	$\Delta_f H°$ $_{(298)}$	Species	$\Delta_f H°$ $_{(298)}$
CH$_3$F	−56.54 ± 0.07	CH$_3$OOH	−30.96 ± 0.67
	−56.62 ± 0.48	CH$_3$CH$_2$OOH	−38.94 ± 0.81
CH$_3$CH$_2$F	−65.42 ± 1.11	CH$_3$CH$_2$CH$_2$OOH	−44.03 ± 0.67
CH$_3$CH$_2$CH$_2$F	−70.24 ± 1.30	CH$_3$OO$^\bullet$	2.37 ± 1.24
CH$_2$F$_2$	−108.07 ± 1.46	CH$_3$CH$_2$OO$^\bullet$	−6.19 ± 0.92
	−107.67 ± 0.48	CH$_3$CH$_2$CH$_2$OO$^\bullet$	−11.35 ± 1.24
CH$_3$CHF$_2$	−120.87 ± 1.62	CH$_4$	−17.81 ± 0.01
CH$_3$CH$_2$CHF$_2$	−125.82 ± 1.65	CH$_3$CH$_3$	−20.05 ± 0.04
CHF$_3$	−166.71 ± 1.97	CH$_3$CH$_2$CH$_3$	−25.01 ± 0.06
	−166.09 ± 0.48	CH$_3$CH$_2$CH$_2$CH$_3$	−30.07 ± 0.08

Tab. 7.1 (continued)

Species	$\Delta_f H°$ (298)	Species	$\Delta_f H°$ (298)
CH_3CF_3	-180.51 ± 2.05^a	$CH_3O^•$	5.15 ± 0.08 [c]
$CH_3CH_2CF_3$	-185.48 ± 2.15^a	$CH_3CH_2O^•$	-3.01^d
$CH_3^•$	34.98 ± 0.02^c	OH	8.96 ± 0.01^c
$CH_3CH_2^•$	28.65 ± 0.07^c	CH_3OH	-47.97 ± 0.04^c
$CH_3CH_2CH_2^•$	24.21 ± 0.24^{gj}	CH_3CH_2OH	-56.07 ± 0.05^i
	24.18^i		
H	52.10^c	$HOO^•$	2.94^{cj}
O	59.57^c	HOOH	-32.39 ± 0.04^{fj}
			-32.37^i
CF_4	-223.15	CHF_2CHF_2 [13]	-212.13
$CF_3CH_2CHF_2$ [13]	-286.18	CF_3CHF_2 [13]	-267.79

7.3.4 Standard Enthalpies

Isodesmic work reactions from M-062x/6-31+g (d,p) method of calculation utilized to perform the calculation of standard enthalpy of formations for tri-, tetra-, and penta-fluorinated ethanols and some radicals are presented in Tab. 7.2. Reference species and their standard enthalpy of formation, along with their uncertainties, have been used in all isodesmic work reactions (Tab. 7.1). The calculated sum of thermal enthalpies using the Gaussian M06-2x/6-31+g (d,p) level of theory for all target fluorinated ethanols, their radicals, and reference species has also been used. The standard enthalpy of formation in kcal mol^{-1} for all reference species is listed in Tab. 7.1. Standard deviation values are listed for all calculated Gaussian M06-2x/6-31+g (d,p) standard enthalpy of formations values. The calculated Gaussian M06-2x/6-31+g (d,p) standard enthalpy of formation for tri-, tetra-, and penta-fluorinated ethanol and some radicals are listed in Tab. 7.2.

Tab. 7.2: Isodesmic reactions used in calculating the standard enthalpy of formation, ΔH°_{rxn} for tri-, tetra-, and penta-fluorinated ethanols and its radicals using the Gaussian M06-2x/6-31+g (d,p) level of theory.

Isodesmic Reactions Target Specie	$\Delta H^\circ_{Rxn\ (298)}$ Hartrees	$\Delta H^\circ_{Rxn\ (298)}$ Kcal/mole[1]	$\Delta_f H^\circ_{(298)}$ kcal mol^{-1}	Error kcal mol^{-1}
CF$_3$CH$_2$OH + CH$_4$ = CH$_3$CH$_2$OH + CHF$_3$ $-$452.696144 $-$40.447961 $-$154.926666 $-$338.204669 $-$17.81 $-$56.21 $-$166.71	0.01277	8.013303	$-$213.1	±2.03
CF$_3$CH$_2$OH + CH$_3$CH$_3$ = CH$_3$CH$_2$OH + CH$_3$CF$_3$ $-$452.696144 $-$79.717768 $-$154.926666 $-$377.492814 $-$20.05 $-$56.21 $-$180.51	$-$0.00557	$-$3.49398	$-$213.2	±2.14
Reported $\Delta_f H^\circ$ (298) kcal mol^{-1}			$-$213.15 ±2.09	
Standard Deviation over rxns 0.05 kcal mol^{-1}				
CF$_3$CH•OH + CH$_3$CH$_3$ = CH$_3$CH$_2$O• + CH$_3$CF$_3$ $-$452.048269 $-$79.717768 $-$154.268107 $-$377.492814 $-$20.05 $-$3.01 $-$180.51	0.005116	3.210341	$-$166.7	± 2.09
CF$_3$CH•OH + CH$_4$ = CH$_3$CH$_2$O• + CHF$_3$ $-$452.048269 $-$40.447961 $-$154.268107 $-$338.204669 $-$17.81 $-$3.01 $-$166.71	0.023454	14.71762	$-$166.6	±1.98
Reported $\Delta_f H^\circ$ (298) kcal mol^{-1}			$-$166.65±2.04	
Standard Deviation over rxns 0.05 kcal mol^{-1}				
CF$_3$CH$_2$O• +CH$_3$CH$_2$CH$_3$= CH$_3$CH$_2$O• + CH$_3$CH$_2$CF$_3$ $-$452.023032 $-$118.99092 $-$154.268107 $-$416.766483 $-$25.02 $-$3.01 $-$185.48	$-$0.02064	$-$12.9537	$-$150.5	±2.21
CF$_3$CH$_2$O• + CH$_4$ = CH$_3$CH$_2$O• + CHF$_3$ $-$452.023032 $-$40.447961 $-$154.268107 $-$338.204669 $-$17.81 $-$3.01 $-$166.71	$-$0.00178	$-$1.11885	$-$150.8	±1.98

Tab. 7.2 (continued)

Isodesmic Reactions Target Specie	$\Delta H°_{Rxn (298)}$ Hartrees	$\Delta H°_{Rxn (298)}$ Kcal/mole[1]	$\Delta_f H°_{(298)}$ kcal mol^{-1}	Error kcal mol^{-1}
Reported $\Delta_f H°$ (298) kcal mol^{-1}			−150.70±2.09	
Standard Deviation over rxns 0.16 kcal mol^{-1}				
$CHF_2CHFOH + CH_3CH_3 = CH_3CH_2OH$ $+ CH_3CF_3$ −452.686514 −79.717768 −154.926666 −377.492814 −20.05 −56.21 −180.51	−0.0152	−9.5369	−207.1	±2.14
$CHF_2CHFOH + CH_4 = CH_3CH_2OH$ $+ CHF_3$ −452.686514 −40.447961 −154.926666 −338.204669 −17.81 −56.21 −166.71	0.00314	1.970381	−207.1	±2.03
Reported $\Delta_f H°$ (298) kcal mol^{-1}			−207.10±2.09	
Standard Deviation over rxns 0.00 kcal mol^{-1}				
$C•F_2CHFOH + CH_4 = CH_3CH_2O•$ $+CHF_3$ −452.032761 −40.447961 −154.268107 −338.204669 −17.81 −3.01 −166.71	0.007946	4.986194	−156.9	±1.98
$C•F_2CHFOH + CH_3CH_3 = CH_3CH_2O• +$ CH_3CF_3 −452.032761 −79.717768 −154.268107 −377.492814 −20.05 −3.01 −180.51	−0.01039	−6.52108	−156.9	±2.09
Reported $\Delta_f H°$ (298) kcal mol^{-1}			−156.90 ±2.04	
Standard Deviation over rxns 0.00 kcal mol^{-1}				
$CHF_2C•FOH + CH_3CH_3 = CH_3CH_2O• +$ CH_3CF_3 −452.028711 −79.717768 −154.268107 −377.492814 −20.05 −5.01 −180.51	−0.01444	−9.0625	−154.4	±2,09

Tab. 7.2 (continued)

Isodesmic Reactions Target Specie	$\Delta H^\circ_{Rxn\,(298)}$ Hartrees	$\Delta H^\circ_{Rxn\,(298)}$ Kcal/mole[1]	$\Delta_f H^\circ_{(298)}$ kcal mol^{-1}	Error kcal mol^{-1}
$CHF_2C\bullet FOH + CH_3CH_2CH_3 = CH_3CH_2O\bullet +$ $CH_3CH_2CF_3$ $-452.028711\ -118.990915\ -154.268107$ -416.766483 $-25.02\ -5.01\ -185.48$	-0.01496	-9.39006	-154.1	± 2.21
Reported $\Delta_f H^\circ$ (298) kcal mol^{-1}			-154.30 ± 2.15	
Standard Deviation over rxns 0.16 kcal mol^{-1}				
$CHF_2CHFO\bullet + CH_4 = CH_3CH_2O\bullet + CHF_3$ $-452.019492\ -40.447961\ -154.268107$ -338.204669 $-17.81\ -3.01$ -166.71	-0.00532	-3.34024	-148.6	± 1.98
$CHF_2CHFO\bullet + CH_3CH_2CH_3 = CH_3CH_2O\bullet +$ $CH_3CH_2CF_3$ $-452.019492\ -118.99092\ -154.268107$ -416.766483 $-25.02\ -3.01\ -185.48$	-0.02418	-15.1751	-148.3	± 2.21
Reported $\Delta_f H^\circ$ (298) kcal mol^{-1}			-148.50 ± 2.09	
Standard Deviation over rxns 0.16 kcal mol^{-1}				
$CH_2FCF_2OH + CH_4 = CH_3CH_2OH + CHF_3$ $-452.696009\ -40.447961\ -154.926666$ -338.204669 $-17.81\ -56.21\ -166.71$	0.012635	7.928589	-213.0	± 2.03
$CH_2FCF_2OH + CH_3CH_3 = CH_3CH_2OH$ $+ CH_3CF_3$ $-452.696009\ -79.717768\ -154.926666$ -377.492814 $-20.05\ -56.21\ -180.51$	-0.0057	-3.57869	-213.1	± 2.14
Reported $\Delta_f H^\circ$ (298) kcal mol^{-1}			-213.05 ± 2.09	
Standard Deviation over rxns 0.05 kcal mol^{-1}				
$C\bullet HFCF_2OH + CH_4 = CH_3CH_2O\bullet + CHF_3$ $-452.036986\ -40.447961\ -154.268107$ -338.204669 $-17.81\ -3.01\ -166.71$	0.012171	7.637424	-159.5	± 1.98

Tab. 7.2 (continued)

Isodesmic Reactions Target Specie	$\Delta H°_{Rxn\ (298)}$ Hartrees	$\Delta H°_{Rxn\ (298)}$ Kcal/mole[1]	$\Delta_f H°_{(298)}$ kcal mol^{-1}	Error kcal mol^{-1}
C•HFCF$_2$OH + CH$_3$CH$_3$ = CH$_3$CH$_2$O•+ CH$_3$CF$_3$ −452.036986 −79.717768 −154.268107 −377.492814 −20.05 −3.01 −180.51	−0.00617	−3.86985	−159.6	±2.09
Reported $\Delta_f H°$ (298) kcal mol^{-1}			−159.55±2.04	
Standard Deviation over rxns 0.08 kcal mol^{-1}				
CH$_2$FCF$_2$O• + CH$_4$ = CH$_3$CH$_2$O• +CHF$_3$ −452.019556 −40.447961 −154.268107 −338.204669 −17.81 −3.01 −166.71	−0.00526	−3.30008	−148.6	±1.98
CH$_2$FCF$_2$O• + CH$_3$CH$_3$ = CH$_3$CH$_2$O•+ CH$_3$CF$_3$ −452.019556 −79.717768 −154.268107 −377.492814 −20.05 −3.01 −180.51	−0.0236	−14.8074	−148.7	±2.09
Reported $\Delta_f H°$ (298) kcal mol^{-1}			−148.65±2.04	
Standard Deviation over rxns 0.05 kcal mol^{-1}				
CF$_3$CFHOH + CH$_4$ = CH$_3$CH$_2$OH+ CF$_4$ −551.950082 −40.447961 −154.926666 −437.46666 −17.81 −56.21 −223.15	0.004717	2.959965	−264.5	±0.10
CF$_3$CFHOH + CH$_3$CH$_3$ = CH$_3$CH$_2$OH + CHF$_2$CHF$_2$ −551.950082 −79.717768 −154.926666 −476.716001 −20.05 −56.21 −212.13	0.025183	15.80258	−264.1	±0.09
Reported $\Delta_f H°$ (298) kcal mol^{-1}			−264.30±0.10	
Standard Deviation over rxns 0.20 kcal mol^{-1}				
CF$_3$C•FOH + CH$_4$ = CH$_3$CH$_2$O•+ CF$_4$ −551.293882 −40.447961 −154.268107 −437.46666 −17.81 −3.01 −223.15	0.007076	4.440261	−212.8	±0.10

Tab. 7.2 (continued)

Isodesmic Reactions Target Specie	$\Delta H°_{Rxn\,(298)}$ Hartrees	$\Delta H°_{Rxn\,(298)}$ Kcal/mole[1]	$\Delta_f H°_{(298)}$ kcal mol^{-1}	Error kcal mol^{-1}
$CF_3C•FOH + CH_3CH_3 = CH_3CH_2O•$ $+ CHF_2CHF_2$ −551.293882 − 79.717768 −154.268107 −476.716001 −20.05 −3.01 −212.13	0.027542	17.28288	−212.4	±0.40
Reported $\Delta_f H°$ (298) kcal mol^{-1}			−212.60±0.30	
Standard Deviation over rxns 0.20 kcal mol^{-1}				
$CF_3CFHO• + CH_4 = CH_3CH_2O•+ CF_4$ −551.281264 −40.447961 −154.268107 −437.46666 −17.81 −3.01 −223.15	−0.00554	−3.47766	−204.9	±0.10
$CF_3CFHO• + CH_3CH_3 = CH_3CH_2O• +$ CHF_2CHF_2 −551.281264 −79.717768 −154.268107 −476.716001 −20.05 −3.01 −212.13	0.014924	9.364959	−204.5	±0.40
Reported $\Delta_f H°$ (298) kcal mol^{-1}			−204.70±0.30	
Standard Deviation over rxns 0.20 kcal mol^{-1}				
$CF_2HCF_2OH + CH_4 = CH_3CH_2OH+ CF_4$ −551.947553 − 40.447961 −154.926666 −437.46666 −17.81 −56.21 −223.15	0.002188	1.372992	−262.9	±0.10
$CF_2HCF_2OH + CH_3CH_3 = CH_3CH_2OH+$ CHF_2CHF_2 −551.947553 −79.717768 −154.926666 −476.716001 −20.05 −56.21 −212.13	0.022654	14.21561	−262.5	±0.10
Reported $\Delta_f H°$ (298) kcal mol^{-1}			−262.70±0.10	
Standard Deviation over rxns 0.20 kcal mol^{-1}				
$C•F_2CF_2OH + CH_4 = CH_3CH_2O• + CF_4$ −551.286788 −40.447961 −154.268107 −437.46666 −17.81 −3.01 −223.15	−1.8E−05	−0.0113	−208.3	±0.10

Tab. 7.2 (continued)

Isodesmic Reactions Target Specie	$\Delta H^\circ_{Rxn\ (298)}$ Hartrees	$\Delta H^\circ_{Rxn\ (298)}$ Kcal/mole[1]	$\Delta_f H^\circ_{(298)}$ kcal mol^{-1}	Error kcal mol^{-1}
C•F$_2$CF$_2$OH + CH$_3$CH$_3$ = CH$_3$CH$_2$O• + CHF$_2$CHF$_2$ −551.286788 −79.717768 −154.268107 −476.716001 −20.05 −3.01 −212.13	0.020448	12.83132	−207.9	±0.40
Reported $\Delta_f H^\circ$ (298) kcal mol^{-1}			−208.10±0.30	
Standard Deviation over rxns 0.20 kcal mol^{-1}				
CF$_2$HCF$_2$O• + CH$_4$ = CH$_3$CH$_2$O• + CF4 −551.26873 −40.447961 −154.268107 −437.46666 −17.81 −3.01 −223.15	−0.01808	−11.3429	−197.0	±0.10
CF$_2$HCF$_2$O• + CH$_3$CH$_3$ = CH$_3$CH$_2$O• + CHF$_2$CHF$_2$ −551.26873 −79.717768 −154.268107 −476.716001 −20.05 −3.01 −212.13	0.00239	1.499749	−196.6	±0.40
Reported $\Delta_f H^\circ$ (298) kcal mol^{-1}			−196.80±0.30	
Standard Deviation over rxns 0.20 kcal mol^{-1}				
CF$_3$CF$_2$OH + CH$_3$CH$_2$CH$_3$= CH$_3$CH$_2$OH +CF$_3$CH$_2$CHF$_2$ −651.207522 −118.99092−154.926666 −615.260935 −25.02 −56.21 −286.18	0.010836	6.799698	−324.2	±0.10
CF$_3$CF$_2$OH + CH$_3$CH$_3$ = CH$_3$CH$_2$OH+ CF$_3$CHF$_2$ −651.207522 −79.717768 −154.926666 −575.976217 −20.05 −56.21 −267.79	0.022407	14.06062	−318.0	±1.70
CF$_3$CF$_2$OH + CH$_4$ = CF$_3$OH+ CH$_3$CHF$_2$ −651.207522 −40.447961 −413.437575−278.225303 −17.81 −218.11 −120.87	−0.00739	−4.64044	−316.5	±1.70

Tab. 7.2 (continued)

Isodesmic Reactions Target Specie	$\Delta H°_{Rxn\,(298)}$ Hartrees	$\Delta H°_{Rxn\,(298)}$ Kcal/mole[1]	$\Delta_f H°_{(298)}$ kcal mol^{-1}	Error kcal mol^{-1}
Reported $\Delta_f H°$ (298) kcal mol^{-1}			-319.60 ± 1.20	
Standard Deviation over rxns 3.30 kcal mol^{-1}				
$CF_3CF_2O\bullet + CH_4 = CHF_2O\bullet + CH_3CF_3$ $-650.527081\ -40.447961\ -313.498636$ -377.492814 $-17.81\ -97.82\ -180.51$	-0.01641	-10.2962	-250.2	±2.06
$CF_3CF_2O\bullet + CH_3CH_2CH_3 = CH_3CH_2O\bullet$ $+CF_3CH_2CHF_2$ $-650.527081\ -118.990915\ -154.268107$ -615.260935 $-25.02\ -3.01\ -286.18$	-0.01105	-6.93148	-257.2	±0.10
$CF_3CF_2O\bullet + CH_3CH_3 = CH_3CH_2O\bullet+$ CF_3CHF_2 $-650.527081\ -79.717768\ -154.268107$ -575.976217 $-20.05\ -3.01\ -267.79$	0.000525	0.329443	-251.1	±1.70
Reported $\Delta_f H°$ (298) kcal mol^{-1}			-252.80 ± 1.30	
Standard Deviation over rxns 3.11 kcal mol^{-1}				

Hartrees, kcal mole^{-1}
*SD Standard Deviation kcal mol^{-1}
Errors reported avg of sum of uncertainties in rxn's reference species

7.3.5 Bond Energies

The calculated BDE values in kcal/mol for methyl C–H, ethyl C–H, and hydroxyl O–H bonds are listed in Tab. 7.3.

The difference in BDEs of products and reactants for hemolysis is called the BDE. The BDE values don't depend on the pathway by which it occurs; it doesn't depend on how bonds are formed or on how bonds break. Therefore, BDEs are state functions. Energetics of chemical processes can be assessed using BDE values.

Tab. 7.3: Bond dissociation energy (BDE) of tri-, tetra-, and penta-fluorinated ethanols using Gaussian M-062x/6-31 + g (d,p) method of calculation.

Reactions	Bond Dissociation Energy[a] (Kcal mol^{-1})	* Error Kcal mol^{-1}
CF$_3$CH–HOH CF$_3$CH–HOH = H• + CF$_3$CH• OH −213.15 ±2.09 **52.10** −166.65±2.04	**98.60± 2.07**	± 2.07
CF$_3$CH$_2$O–H CF$_3$CH$_2$O–H = H• + CF$_3$CH$_2$O• −213.15 ±2.09 **52.10** −150.70±2.09	**114.55±2.09**	±2.09
C–HF$_2$CHFOH C–HF$_2$CHFOH = H• + C• F$_2$CHFOH −207.10±2.09 **52.10** −156.90 ±2.04	**102.30±2.07**	±2.07
CHF$_2$C–HFOH CHF$_2$C–HFOH = H• + CHF$_2$C• FOH −207.10±2.09 **52.10** −154.30±2.15	**104.90±2.10**	±2.10
CHF$_2$CHFO–H CHF$_2$CHFO–H = H• + CHF$_2$CHFO• −207.10±2.09 **52.10** −148.50±2.09	**110.70±2.09**	±2.09
HC–HFCF$_2$OH HC–HFCF$_2$OH = H• + C• HFCF$_2$OH −213.05±2.09 **52.10** −159.55±2.04	**105.60±2.07**	±2.07
CH$_2$FCF$_2$O–H CH$_2$FCF$_2$O–H= H• + CH$_2$FCF$_2$O• −213.05±2.09 **52.10** −148.65±2.04	**116.50±2.07**	±2.07
CF$_3$CF–HOH CF$_3$CF–HOH= H• + CF$_3$CF• OH −264.30±0.10 **52.10** −212.60±0.30	**103.80±0.20**	±0.20
CF$_3$CFHO–H CF$_3$CFHO–H= H• + CF$_3$CFHO• −264.30±0.10 **52.10** −204.70±0.30	**111.70±0.20**	±0.20
CF$_2$–HCF$_2$OH CF$_2$–HCF$_2$OH= H• + CF$_2$• CF$_2$OH −262.70±0.10 **52.10** −208.10±0.30	**106.70±0.20**	±0.20
CF$_2$HCF$_2$O–H CF$_2$HCF$_2$O–H= H• + CF$_2$HCF$_2$O• −262.70±0.10 **52.10** −196.80±0.30	**118.00±0.20**	±0.20
CF$_3$CF$_2$O–H CF$_3$CF$_2$O–H= H• + **CF$_3$CF$_2$O•** −319.60±1.20 **52.10** −252.80±1.30	**118.90±1.30**	±1.30

*Errors reported avg of sum of uncertainties in rxn's reference specie

7.4 Optimized Structures

The GaussView software has been utilized along with Gaussian output files to provide a picture of the Gaussian M06-2x/6-31 + g (d,p) optimized structure for each molecule in this chapter (Tab. 7.4).

Tab. 7.4: Optimized geometries for target tri-, tetra-, and penta-fluorinated ethanols and their related radicals calculated by Gaussian M06-2x/6-31 + g (d,p) level of theory.

CF_3CH_2OH

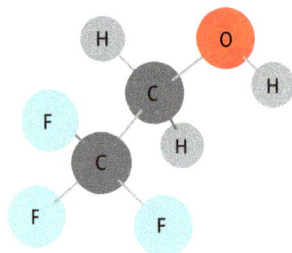

$CF_3CH \cdot OH$

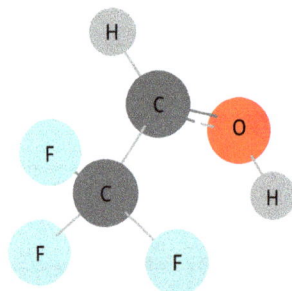

$CF_3CH_2O \cdot$

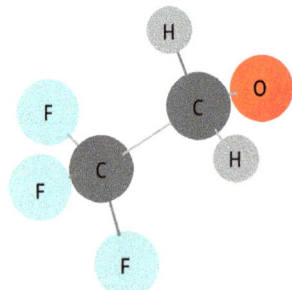

Tab. 7.4 (continued)

CHF$_2$CHFOH

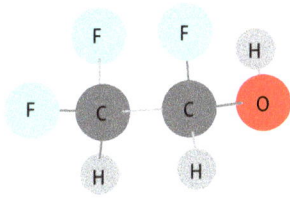

C•F$_2$CHFOH

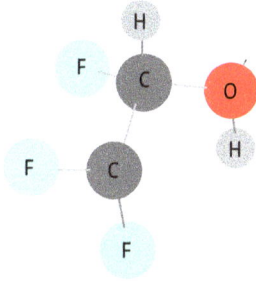

CHF$_2$C•FOH

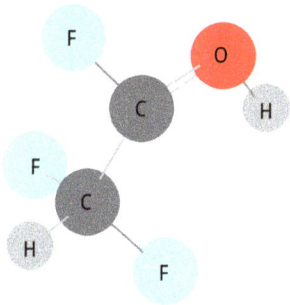

CHF$_2$CHFO•

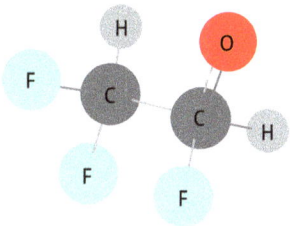

Tab. 7.4 (continued)

CH₂FCF₂OH

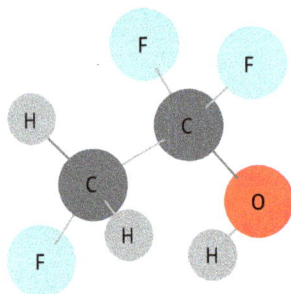

C•HFCF₂OH

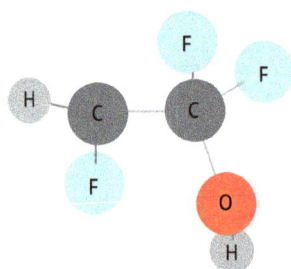

CH₂FCF₂O•

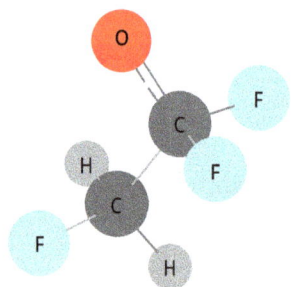

CF₃CFHOH

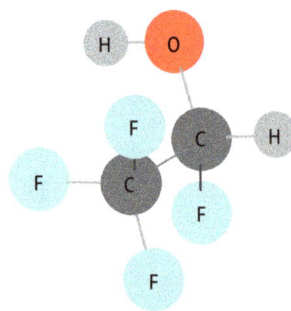

Tab. 7.4 (continued)

CF₃CF•OH

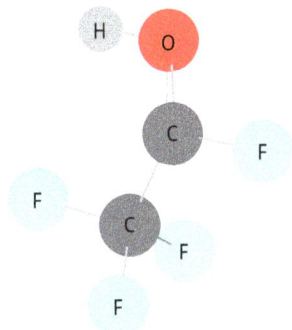

CF₃CFHO•

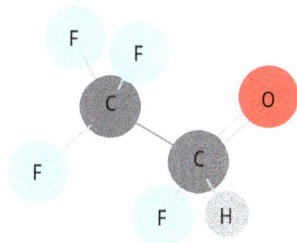

CF₂HCF₂OH

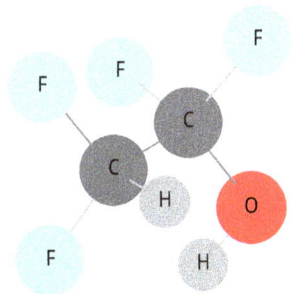

C•F₂CF₂OH

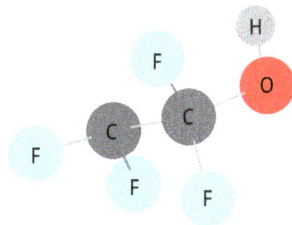

Tab. 7.4 (continued)

CF$_2$HCF$_2$O•

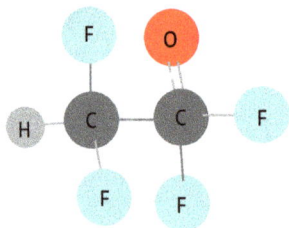

CF$_3$CF$_2$OH

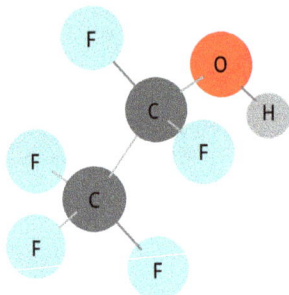

CF$_3$CF$_2$O•

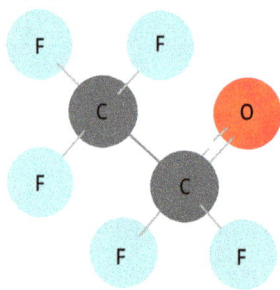

7.5 Cartesian Coordinates

The Cartesian coordinates for target tri-, tetra-, and penta-fluorinated ethanols and their related radicals at the Gaussian M-062x/6-31 + g (d,p) level of theory are listed in Tab. 7.5.

Tab. 7.5: Cartesian coordinates in angstroms for target fluorinated ethanol and their related radical's geometries at the M-062x/6-31+g (d,p) level of theory.

Molecule	Cartesian Coordinates

CF₃CH₂OH

Center Number	Atomic Number	Forces (Hartrees/Bohr) X	Y	Z
1	6	0.000015792	0.000045146	0.000015456
2	6	0.000056272	-0.000131331	-0.000001091
3	9	-0.000036018	-0.000005130	-0.000021513
4	9	0.000035235	0.000054970	-0.000018991
5	9	-0.000001701	0.000051428	0.000034529
6	1	-0.000016200	0.000007968	0.000014306
7	1	-0.000023806	-0.000003074	-0.000011946
8	8	-0.000022602	-0.000015046	0.000001930
9	1	-0.000007275	-0.000024931	-0.000012675

CF₃C•HOH

Center Number	Atomic Number	Forces (Hartrees/Bohr) X	Y	Z
1	6	-0.000055227	-0.000030630	-0.000017226
2	6	0.000117503	0.000159657	-0.000153816
3	9	-0.000061675	0.000035455	0.000071731
4	9	-0.000053827	-0.000176074	0.000047967
5	9	0.000080802	-0.000106445	0.000024638
6	1	0.000001337	0.000046611	-0.000002493
7	8	0.000004111	0.000030529	-0.000059199
8	1	-0.000033025	0.000040898	0.000068398

CF₃CH₂O•

Center Number	Atomic Number	Forces (Hartrees/Bohr) X	Y	Z
1	6	-0.000037529	0.000088167	0.000122350
2	6	-0.000033636	-0.000121335	-0.000217962
3	9	0.000020892	0.000000631	0.000073365
4	9	0.000016383	-0.000001502	0.000002646
5	9	-0.000021622	0.000041297	0.000066627
6	1	-0.000020139	-0.000044741	-0.000019536
7	1	-0.000000403	-0.000003543	-0.000012733
8	8	0.000076054	0.000041026	-0.000014758

CHF₂CHFOH

Center Number	Atomic Number	Forces (Hartrees/Bohr) X	Y	Z
1	6	0.000034684	-0.000097837	0.000016859
2	6	0.000077734	-0.000008144	0.000008869
3	9	-0.000027798	0.000012351	0.000007573
4	9	-0.000002203	0.000054383	-0.000025902
5	1	-0.000018584	0.000015264	0.000009448
6	9	0.000005269	0.000001678	-0.000006105
7	9	-0.000005043	-0.000008859	0.000006453
8	8	-0.000067666	0.000036885	-0.000008025
9	1	0.000003607	-0.000005722	-0.000009170

C•F₂CHFOH

Center Number	Atomic Number	Forces (Hartrees/Bohr) X	Y	Z
1	6	-0.000071214	0.000095917	0.000002631
2	6	-0.000097297	-0.000427988	0.000070858
3	9	0.000109064	0.000106013	0.000068648
4	9	0.000002068	0.000252877	-0.000098474
5	1	-0.000001211	0.000016670	-0.000018084
6	9	0.000033735	-0.000023289	-0.000027474
7	8	0.000054362	-0.000036847	0.000018777
8	1	-0.000029507	0.000016646	-0.000016882

Tab. 7.5 (continued)

Molecule	Cartesian Coordinates
CHF$_2$C•FOH	

Center Number	Atomic Number	Forces (Hartrees/Bohr) X	Y	Z
1	6	-0.000038782	0.000051546	0.000051509
2	6	-0.000109395	-0.000076573	-0.000071886
3	9	0.000040508	0.000010628	0.000077586
4	9	0.000020881	0.000006702	-0.000016053
5	1	0.000007533	-0.000016674	-0.000041204
6	9	0.000090694	-0.000038267	-0.000029261
7	8	-0.000040547	0.000056005	0.000004711
8	1	0.000029108	0.000006632	0.000024597

CHF$_2$CHFO•

Center Number	Atomic Number	Forces (Hartrees/Bohr) X	Y	Z
1	6	0.000055628	-0.000106091	0.000003381
2	6	0.000053463	-0.000000718	-0.000007006
3	9	-0.000043989	-0.000020986	0.000005782
4	9	-0.000016210	-0.000035868	0.000004447
5	1	-0.000011020	-0.000013533	0.000007886
6	1	-0.000010458	0.000024649	-0.000005524
7	9	-0.000051835	0.000115358	-0.000005383
8	8	0.000022121	0.000009470	-0.000053144
9	1	0.000002301	0.000027718	0.000049560

CH$_2$FCF$_2$OH

Center Number	Atomic Number	Forces (Hartrees/Bohr) X	Y	Z
1	6	0.000135651	0.000057453	0.000051033
2	6	-0.000243304	0.000078716	-0.000013330
3	9	-0.000116433	0.000005813	0.000097761
4	1	-0.000077576	-0.000009293	0.000043932
5	1	-0.000038364	-0.000018670	0.000006345
6	9	0.000035849	0.000027631	-0.000096988
7	9	0.000045365	0.000036806	-0.000102497
8	8	0.000181449	-0.000179561	-0.000003213
9	1	0.000077364	0.000001106	0.000016957

C•HFCF$_2$OH

Center Number	Atomic Number	Forces (Hartrees/Bohr) X	Y	Z
1	6	-0.000009430	0.000025594	0.000005992
2	6	0.000017955	-0.000004162	0.000000569
3	9	0.000001322	0.000052928	-0.000001501
4	1	-0.000003779	0.000024041	0.000008036
5	9	-0.000012190	-0.000046066	0.000011421
6	9	-0.000025702	-0.000032617	0.000028822
7	8	0.000010500	-0.000018093	-0.000017353
8	1	0.000021325	-0.000001626	-0.000035986

CH$_2$FCF$_2$O•

Center Number	Atomic Number	Forces (Hartrees/Bohr) X	Y	Z
1	6	0.000157230	-0.000033765	0.000067368
2	6	-0.000084968	-0.000038414	-0.000127598
3	9	-0.000060007	0.000028432	0.000052860
4	1	-0.000002632	-0.000019789	-0.000013222
5	1	-0.000027022	0.000016735	-0.000013331
6	9	-0.000031881	-0.000031453	0.000013157
7	9	0.000080040	0.000071255	0.000005918
8	8	-0.000030759	0.000006999	0.000014849

Tab. 7.5 (continued)

Molecule	Cartesian Coordinates

CF₃CFHOH

Center Number	Atomic Number	Forces (Hartrees/Bohr)		
		X	Y	Z
1	6	-0.000139743	-0.000019177	-0.000028107
2	6	-0.000044583	-0.000023229	-0.000002049
3	9	0.000017824	0.000030702	0.000057967
4	9	-0.000069952	-0.000066058	-0.000012198
5	9	0.000058051	-0.000059520	-0.000018868
6	9	0.000166545	0.000025560	-0.000007913
7	1	0.000003967	0.000022634	0.000010499
8	8	-0.000043206	0.000028605	-0.000066782
9	1	0.000051097	0.000060482	0.000067450

CF₃C•FOH

Center Number	Atomic Number	Forces (Hartrees/Bohr)		
		X	Y	Z
1	6	0.000033580	0.000004460	0.000005322
2	6	0.000020098	0.000029526	-0.000031153
3	9	0.000021581	-0.000000752	0.000016405
4	9	0.000016549	0.000006426	0.000007990
5	9	-0.000041946	0.000031215	0.000004491
6	9	-0.000000439	-0.000039185	-0.000010685
7	8	-0.000042830	-0.000023720	0.000011783
8	1	-0.000006593	-0.000007971	-0.000004153

CF₃CFHO•

Center Number	Atomic Number	Forces (Hartrees/Bohr)		
		X	Y	Z
1	6	0.000536610	0.000642487	-0.000238587
2	6	-0.000336421	-0.000010459	-0.000050530
3	9	0.000179754	0.000022776	0.000069636
4	9	0.000013212	-0.000037718	0.000038270
5	9	0.000033732	0.000056583	-0.000043087
6	9	-0.000277185	-0.000172926	0.000053380
7	1	-0.000020915	-0.000096792	0.000011479
8	8	-0.000128787	-0.000403952	0.000159440

CF₂HCF₂OH

Center Number	Atomic Number	Forces (Hartrees/Bohr)		
		X	Y	Z
1	6	0.000034684	-0.000097837	0.000016859
2	6	0.000077734	-0.000008144	0.000008869
3	9	-0.000027798	0.000012351	0.000007573
4	9	-0.000002203	0.000054383	-0.000025902
5	1	-0.000018584	0.000015264	0.000009448
6	9	0.000005269	0.000001678	-0.000006105
7	9	-0.000005043	-0.000008859	0.000006453
8	8	-0.000067666	0.000036885	-0.000008025
9	1	0.000003607	-0.000005722	-0.000009170

Tab. 7.5 (continued)

Molecule	Cartesian Coordinates		

C•F$_2$CF$_2$OH

Center Number	Atomic Number	Forces (Hartrees/Bohr)		
		X	Y	Z
1	6	0.000024554	0.000023392	0.000004415
2	6	-0.000034477	0.000024861	-0.000019704
3	9	0.000055726	0.000012426	0.000032150
4	9	0.000019915	0.000028025	-0.000017047
5	9	0.000004005	-0.000011188	-0.000003720
6	9	0.000015833	-0.000043273	0.000042429
7	8	-0.000068386	-0.000006654	0.000010769
8	1	-0.000017169	-0.000027587	-0.000049293

CF$_2$HCF$_2$O•

Center Number	Atomic Number	Forces (Hartrees/Bohr)		
		X	Y	Z
1	6	0.000006128	-0.000032081	0.000064624
2	6	0.000009899	-0.000032233	-0.000045587
3	9	-0.000037818	-0.000059393	0.000013603
4	9	0.000041223	0.000037095	-0.000050906
5	1	-0.000021281	0.000006425	-0.000018170
6	9	-0.000013457	0.000031071	0.000014828
7	9	0.000079466	0.000042020	-0.000014271
8	8	-0.000064160	0.000007097	0.000035879

CF$_3$CF$_2$OH

Center Number	Atomic Number	Forces (Hartrees/Bohr)		
		X	Y	Z
1	6	-0.000029279	0.000029059	-0.000014751
2	6	0.000037295	0.000003602	0.000009039
3	9	-0.000004027	-0.000031190	-0.000022595
4	9	-0.000016447	-0.000026160	-0.000006065
5	9	0.000008918	-0.000023108	0.000015387
6	9	-0.000006450	0.000029260	-0.000000832
7	9	0.000001454	0.000001809	-0.000025482
8	8	-0.000002384	0.000003392	0.000022090
9	1	0.000010920	0.000013336	0.000023209

CF$_3$CF$_2$O•

Center Number	Atomic Number	Forces (Hartrees/Bohr)		
		X	Y	Z
1	6	0.000090683	0.000175823	0.000176247
2	6	-0.000004772	-0.000005489	-0.000074419
3	9	-0.000035079	0.000031055	-0.000015124
4	9	-0.000026035	-0.000069252	0.000016300
5	9	0.000038428	-0.000012873	-0.000000972
6	9	-0.000080090	-0.000021765	-0.000033887
7	9	0.000031147	-0.000065731	-0.000034377
8	8	-0.000014283	-0.000031768	-0.000033768

7.6 Conclusion

The ab initio and Global-hybrid meta-GGA density function method, Gaussian M06-2*x*/6-31 + g (d,p), used to calculate thermodynamic properties of 18 tri-, tetra-, and penta-fluorinated ethanols and their related radicals are presented. Calculated standard enthalpy of formation at the Gaussian M06-2*x*/6-31 + g (d,p) using multiple work reactions are also presented; work reactions were employed for cancellation of calculation errors (Tab. 7.2).

The calculated thermochemical properties: the O–H, secondary methyl C–H, and ethyl C–H BDEs, and standard enthalpy of formation (298 K) values for tri-, tetra-, and penta-fluorinated ethanols and their related radicals: $CH_3CH_2–xFxOH$, $CH_3–xFxCH_2–xFxOH$, and $CH_2–xFxCH_2OH$ are tabulated in Tab. 7.3.

The C–H BDEs ranged from 98.6 to 104.9 kcal mol^{-1} on the secondary ethyl carbons and from 103.3 to 106.7 kcal mol^{-1} on the primary methyl carbons. The O–H bond energies range from 110.7 to 118.9 kcal mol^{-1}. The calculated O–H bond energies increased by intruding more fluorine atoms into either methyl or ethyl carbons.

Chapter Questions

1) What's the aim of the chapter?
2) What are the uses for fluorinated alcohols?
3) List names and formulas of the fluorinated alcohols discussed in this chapter?
4) What's the method used in calculating thermodynamic properties discussed in the chapter?
5) What is the name and structure of the fluorinated found to have the highest standard enthalpy of formation? And which is found to have the highest C–H and O–H bond dissociation energy values?

References

[1]   An X., Xiao J., Fluorinated Alcohols: Magic Reaction Medium and Promoters for Organic Synthesis. The Chemical record, a journal of the chemical society of Japan. 2020, 20(2): 142–161.
[2]   Zhang M., Peyear T., Patmanidis I.,Greathouse D.V., Marrink S.J., Andersen O.S., and Ingólfsson H.I. Fluorinated Alcohols' Effects on Lipid Bilayer Properties. Biophysical Journal. 2018, 115(4): 679–689.
[3]   J. P. Ducrotoy, K.Mazik. Chemical Introductions to the Systems: Point Source Pollution (Persistent Chemicals). Treatise on Estuarine and Coastal Science. 2011, Volume 8: 71–111
[4]   Bao G, Abe RY, Akutsu Y. Bond dissociation energy and thermal stability of energetic materials. Journal of Thermal Analysis and Calorimetry. 2021 Mar;143(5):3439–45.
[5]   Sorensen JJ, Tieu E, Morse MD. Bond dissociation energies of lanthanide sulfides and selenides. The Journal of Chemical Physics. 2021 Mar 28;154(12):124307.

[6]   https://chem.libretexts.org/Bookshelves/Organic_Chemistry/Organic_Chemistry_(McMurry)/06%
      3A_An_Overview_of_Organic_Reactions/6.08%
      3A_Describing_a_Reaction__Bond__Dissociation_Energies.

[7]   https://chem.libretexts.org/Courses/Heartland_Community_College/HCC%3A_Chem_161/5%3A_Ther
      mochemistry/5.7%3A_Enthalpy_of_Formation

[8]   Gammon, Ebbing, General Chemistry 11th edition, Boston MA, Cengage Learning, 2017, pp 608–610.
      https://en.wikipedia.org/wiki/Radical_(chemistry)

[9]   Wang H., Castillo A., Bozzelli J.W., Thermochemical properties enthalpy, entropy, and heat capacity
      of C1-4 fluorinated hydrocarbons. J. Phys. Chem. 2015.

[10] Schneider, W. F.; Wallington, T. J. Ab Initio Investigation of the Heats of Formation of Several
      Trifluoromethyl Compounds. J. Phys. Chem. 1993, 97, 12783–12788.

[11] Expanding The Limits of Computational Chemistry, 2015–2019, *Density Function Method*, 7/24/2020, '
      https://gaussian.com/dft/.

[12] Wang, H.; Castillo, Á.; Bozzelli, J. W. Thermochemical Properties Enthalpy, Entropy, and Heat
      Capacity of C1–C4 Fluorinated Hydrocarbons: Fluorocarbon Group Additivity. J. Phys. Chem. A 2015,
      119, 8202–8215.

[13] Math is Fun Advanced, 2017, *Standard Deviation and Variance*, 12/2019,{https://www.mathsisfun.com/
      data/standard-deviation.html

[14] Wang H., Bozzelli J. W., Thermochemical properties and Bond Dissociation Energy for Fluorinated
      Methanol and fluorinated methyl hydroperoxides,J. Phys. Chem. 2016.

[15] Ruscic, B., Active Thermochemical Tables: Sequential Bond Dissociation Enthalpy of Methane,
      Ethane, and Methanol and related Thermochemistry. J. Phys. Chem. A 2015, 119, 7810–7837.

[16] Burke, S. M.; Simmie, J. M.; Curran, H. J. Critical Evaluation of Thermochemical Properties of C1–C4
      Species: Updated Group Contributions to Estimate Thermochemical Properties. J. Phys. Chem. Ref.
      Data 2015, 44, 013101.

[17] Chase, M. W. J. NIST-JANAF Thermochemical Tables. J. Phys. Chem. Ref. Data. 1998, Monograph 9,
      1–1951.

[18] Luo, X.; Fleming, P. R.; Rizzo, T. R. Vibrational Overtone Spectroscopy of the 4 νOH+νOH'
      Combination Level of HOOH via Sequential Local Mode –local Mode Excitation. J. Chem. Phys. 1992,
      96, 5659–5667.

[19] Bodi, A., Kercher, J. P., Bond, C., Meteesatien, P., Sztáray, B., Baer, T., Photoion Photoelectron
      Coincidence Spectroscopy of Primary Amines RCH2NH2 (R = H, CH3, C2H5, C3H7, i-C3H7): Alkylamine
      and Alkyl Radical Heats of Formation by Isodesmic Reaction Networks, 2006, J. Phys. Chem. A,
      doi: 10.1021/jp064739s

[20] C3H7): Alkylamine and Alkyl Radical Heats of Formation by Isodesmic Reaction Networks. J. Phys.
      Chem. A 2006, 110, 13425–13433.

[21] Wang, H.; Bozzelli, J. W. Thermochemical Properties (ΔH_f (298 K), S (298 K), C p(T)) and Bond
      Dissociation Energies for C1–C4 Normal Hydroperoxides and Peroxy Radicals. J. Chem. Eng. Data
      2016, 61, 1836–1849.

[22] Csontos, J.; Rolik, Z.; Das, S.; Kallay, M. High-Accuracy Thermochemistry of Atmospherically
      Important Fluorinated and Chlorinated Methane Derivatives. J. Phys. Chem. A 2010, 114,
      13093–13103.

[23] Gronert, S., *J. Org. Chem.* 2006, 13, 1209.

Disclaimer

Chapter 1: This chapter is another version of the article published by the same author(s) in the following journal: *International Journal of Innovation Scientific Research and Review*, 4: 2404–2408, 2022.

Chapter 2: This chapter is another version of the article published by the same author(s) in the following journal: *American Journal of Applied and Industrial Chemistry*, 6: 13–19, 2022.

Chapter 3: This chapter is another version of the article published by the same author(s) in the following journal: *American Journal of Applied and Industrial Chemistry*, 6(2): 31–35, 2022.

Chapter 4: This chapter is another version of the article published by the same author(s) in the following journal: *Journal of Pharmaceutics and Pharmacology Research*, 5(8), 2022.

Chapter 5: This chapter is another version of the article published by the same author(s) in the following journal: *Journal of Chemistry, Education Research, and Practice*, 5: 127–130, 2021.

Chapter 6: This chapter is another version of the article published by the same author(s) in the following journal: *Journal of Pharmacognosy and Phytochemistry*, 11(4): 295–302, 2022.

Chapter 7: This chapter is another version of the article published by the same author(s) in the following journal: *American Journal of Physical Chemistry*, 11: 32–44, 2022.

https://doi.org/10.1515/9783111316864-008

Appendix: The Periodic Table of Elements

All known elements that make all existing chemical compounds with varying chemical and physical properties discussed in detail in most chapters are arranged by their atomic numbers and listed in the periodic table. The atomic number is the number of protons in the nucleus. The periodic table is the best tool in science, as many chemical and physical properties can be known based on their position in the periodic table. Periodic tables can be obtained using the internet; a good source of information would be https://ptable.com/#Properties. Most periodic tables contain the atomic mass of each element, which is the number of neutrons and protons in each nucleus. Table 1 lists atomic masses for most elements on the periodic table. Some elements have atomic masses that are known very precisely, to the accuracy of several decimal places; other elements, like radioactive elements, have atomic masses that aren't known precisely; and elements with existing isotopes have varying atomic masses depending on how isotopes are isolated.

https://doi.org/10.1515/9783111316864-009

Legend:

Atomic number
Symbol
Name
Weight

C	Solid
Hg	Liquid
H	Gas
Rf	Unknown

Metals
- Alkali metals
- Alkaline earth metals
- Lanthanoids
- Actinoids
- Transition metals
- Post-transition metals

Metalloids

Nonmetals
- Reactive nonmetals
- Noble gases

Chalcogens (group 16), Halogens (group 17)

For elements with no stable isotopes, the mass number of the isotope with the longest half-life is in parentheses.

1	2	3	4	5	6	7	8	9	10	11	12	13	14	15	16	17	18
1 H Hydrogen 1.008																	2 He Helium 4.0026
3 Li Lithium 6.94	4 Be Beryllium 9.0122											5 B Boron 10.81	6 C Carbon 12.011	7 N Nitrogen 14.007	8 O Oxygen 15.999	9 F Fluorine 18.998	10 Ne Neon 20.180
11 Na Sodium 22.990	12 Mg Magnesium 24.305											13 Al Aluminium 26.982	14 Si Silicon 28.085	15 P Phosphorus 30.974	16 S Sulfur 32.06	17 Cl Chlorine 35.45	18 Ar Argon 39.948
19 K Potassium 39.098	20 Ca Calcium 40.078	21 Sc Scandium 44.956	22 Ti Titanium 47.867	23 V Vanadium 50.942	24 Cr Chromium 51.996	25 Mn Manganese 54.938	26 Fe Iron 55.845	27 Co Cobalt 58.933	28 Ni Nickel 58.693	29 Cu Copper 63.546	30 Zn Zinc 65.38	31 Ga Gallium 69.723	32 Ge Germanium 72.630	33 As Arsenic 74.922	34 Se Selenium 78.971	35 Br Bromine 79.904	36 Kr Krypton 83.798
37 Rb Rubidium 85.468	38 Sr Strontium 87.62	39 Y Yttrium 88.906	40 Zr Zirconium 91.224	41 Nb Niobium 92.906	42 Mo Molybdenum 95.95	43 Tc Technetium (98)	44 Ru Ruthenium 101.07	45 Rh Rhodium 102.91	46 Pd Palladium 106.42	47 Ag Silver 107.87	48 Cd Cadmium 112.41	49 In Indium 114.82	50 Sn Tin 118.71	51 Sb Antimony 121.76	52 Te Tellurium 127.60	53 I Iodine 126.90	54 Xe Xenon 131.29
55 Cs Caesium 132.91	56 Ba Barium 137.33	57–71	72 Hf Hafnium 178.49	73 Ta Tantalum 180.95	74 W Tungsten 183.84	75 Re Rhenium 186.21	76 Os Osmium 190.23	77 Ir Iridium 192.22	78 Pt Platinum 195.08	79 Au Gold 196.97	80 Hg Mercury 200.59	81 Tl Thallium 204.38	82 Pb Lead 207.2	83 Bi Bismuth 208.98	84 Po Polonium (209)	85 At Astatine (210)	86 Rn Radon (222)
87 Fr Francium (223)	88 Ra Radium (226)	89–103	104 Rf Rutherfordium (267)	105 Db Dubnium (268)	106 Sg Seaborgium (269)	107 Bh Bohrium (270)	108 Hs Hassium (277)	109 Mt Meitnerium (278)	110 Ds Darmstadtium (281)	111 Rg Roentgenium (282)	112 Cn Copernicium (285)	113 Nh Nihonium (286)	114 Fl Flerovium (289)	115 Mc Moscovium (290)	116 Lv Livermorium (293)	117 Ts Tennessine (294)	118 Og Oganesson (294)

Lanthanoids (6):

57 La Lanthanum 138.91	58 Ce Cerium 140.12	59 Pr Praseodymium 140.91	60 Nd Neodymium 144.24	61 Pm Promethium (145)	62 Sm Samarium 150.36	63 Eu Europium 151.96	64 Gd Gadolinium 157.25	65 Tb Terbium 158.93	66 Dy Dysprosium 162.50	67 Ho Holmium 164.93	68 Er Erbium 167.26	69 Tm Thulium 168.93	70 Yb Ytterbium 173.05	71 Lu Lutetium 174.97

Actinoids (7):

89 Ac Actinium (227)	90 Th Thorium 232.04	91 Pa Protactinium 231.04	92 U Uranium 238.03	93 Np Neptunium (237)	94 Pu Plutonium (244)	95 Am Americium (243)	96 Cm Curium (247)	97 Bk Berkelium (247)	98 Cf Californium (251)	99 Es Einsteinium (252)	100 Fm Fermium (257)	101 Md Mendelevium (258)	102 No Nobelium (259)	103 Lr Lawrencium (266)

Table 1: Common information of elements of the periodic table.

Name	Atomic Symbol	Atomic Number	Atomic Mass	Footnotes
actinium*	Ac	89		
aluminum	Al	13	26.9815386(8)	
americium*	Am	95		
antimony	Sb	51	121.760(1)	g
argon	Ar	18	39.948(1)	g, r
arsenic	As	33	74.92160(2)	
astatine*	At	85		
barium	Ba	56	137.327(7)	
berkelium*	Bk	97		
beryllium	Be	4	9.012182(3)	
bismuth	Bi	83	208.98040(1)	
bohrium*	Bh	107		
boron	B	5	10.811(7)	g, m, r
bromine	Br	35	79.904(1)	
cadmium	Cd	48	112.411(8)	g
caesium (cesium)	Cs	55	132.9054519(2)	
calcium	Ca	20	40.078(4)	g
californium*	Cf	98		
carbon	C	6	12.0107(8)	g, r
cerium	Ce	58	140.116(1)	g
chlorine	Cl	17	35.453(2)	g, m, r
chromium	Cr	24	51.9961(6)	
cobalt	Co	27	58.933195(5)	
copernicium*	Cn	112		
copper	Cu	29	63.546(3)	r
curium*	Cm	96		
darmstadtium*	Ds	110		
dubnium*	Db	105		

Table 1 (continued)

dysprosium	Dy	66	162.500(1)	g
einsteinium*	Es	99		
erbium	Er	68	167.259(3)	g
europium	Eu	63	151.964(1)	g
fermium*	Fm	100		
fluorine	F	9	18.9984032(5)	
francium*	Fr	87		
gadolinium	Gd	64	157.25(3)	g
gallium	Ga	31	69.723(1)	
germanium	Ge	32	72.64(1)	
gold	Au	79	196.966569(4)	
hafnium	Hf	72	178.49(2)	
hassium*	Hs	108		
helium	He	2	4.002602(2)	g, r
holmium	Ho	67	164.93032(2)	
hydrogen	H	1	1.00794(7)	g, m, r
indium	In	49	114.818(3)	
iodine	I	53	126.90447(3)	
iridium	Ir	77	192.217(3)	
iron	Fe	26	55.845(2)	
krypton	Kr	36	83.798(2)	g, m
lanthanum	La	57	138.90547(7)	g
lawrencium*	Lr	103		
lead	Pb	82	207.2(1)	g, r
lithium	Li	3	[6.941(2)]†	g, m, r
lutetium	Lu	71	174.967(1)	g
magnesium	Mg	12	24.3050(6)	
manganese	Mn	25	54.938045(5)	
meitnerium*	Mt	109		
mendelevium*	Md	101		

Table 1 (continued)

mercury	Hg	80	200.59(2)	
molybdenum	Mo	42	95.94(2)	g
neodymium	Nd	60	144.242(3)	g
neon	Ne	10	20.1797(6)	g, m
neptunium*	Np	93		
nickel	Ni	28	58.6934(2)	
niobium	Nb	41	92.90638(2)	
nitrogen	N	7	14.0067(2)	g, r
nobelium*	No	102		
osmium	Os	76	190.23(3)	g
oxygen	O	8	15.9994(3)	g, r
palladium	Pd	46	106.42(1)	g
phosphorus	P	15	30.973762(2)	
platinum	Pt	78	195.084(9)	
plutonium*	Pu	94		
polonium*	Po	84		
potassium	K	19	39.0983(1)	
praseodymium	Pr	59	140.90765(2)	
promethium*	Pm	61		
protactinium*	Pa	91	231.03588(2)	
radium*	Ra	88		
radon*	Rn	86		
roentgenium*	Rg	111		
rhenium	Re	75	186.207(1)	
rhodium	Rh	45	102.90550(2)	
rubidium	Rb	37	85.4678(3)	g
ruthenium	Ru	44	101.07(2)	g
rutherfordium*	Rf	104		
samarium	Sm	62	150.36(2)	g

Table 1 (continued)

zirconium	Zr	40	91.224(2)	g
*Element has no stable nuclides. However, three such elements (Th, Pa, and U) have a characteristic terrestrial isotopic composition, and for these an atomic mass is tabulated.				
†Commercially available Li materials have atomic weights that range between 6.939 and 6.996; if a more accurate value is required, it must be determined for the specific material.				
g Geological specimens are known in which the element has an isotopic composition outside the limits for normal material. The difference between the atomic mass of the element in such specimens and that given in the table may exceed the stated uncertainty.				
m Modified isotopic compositions may be found in commercially available material because it has been subjected to an undisclosed or inadvertent isotopic fractionation. Substantial deviations in the atomic mass of the element from that given in the table can occur.				
r Range in isotopic composition of normal terrestrial material prevents a more precise $Ar(E)$ being given; the tabulated $Ar(E)$ value and uncertainty should be applicable to normal material.				

Source: Adapted from *Pure and Applied Chemistry* 78, no. 11 (2005): 2051–66. © IUPAC (International Union of Pure and Applied Chemistry).

Index

https://doi.org/10.1515/9783111316864-010

www.ingramcontent.com/pod-product-compliance
Lightning Source LLC
Chambersburg PA
CBHW081550220326
41598CB00036B/6627